Organismal Biology: A Laboratory Introduction
Second Edition

J. L. Riopel • Kristen Curran
University of Virginia

KENDALL/HUNT PUBLISHING COMPANY
4050 Westmark Drive   Dubuque, Iowa 52002

Contents

Biology 204
Introduction to Biology Laboratory
(2 credit hours)

Course Information and Policies

The purpose of the Biology 204 lab is to offer the student first-hand experience in the observation and study of organisms. We will examine life forms from simple to complex and we will see both the diversity and the unique properties of living organisms.

With limited time, these studies serve only as an introduction to the complexities and infinite varieties of plants and animals. Only a few organisms have been selected. Study them thoroughly as individual examples exemplifying distinctive features of certain groups. But equally important, remember that these organisms have significant roles in highly complex functioning ecosystems. By continued reading and inquiry, find out what they do.

We will also review the evolution of organisms, when evolution began, and how time and the factors of evolution worked to bring us to the present. Even more importantly, we hope these studies contribute to a framework of information for a life-long interest and appreciation of living organisms.

Laboratory Rules

- Smoking, eating, and drinking are not permitted.

- Waste solids are to be put in designated waste cans, not sinks.

- Maintain a clean work area and return of all supplies to their appropriate location.

- Clean up all broken glass and spills immediately. Place all broken glass and used disposable glass items in the special "GLASS WASTE" can at the front of the room—NOT in the regular trash cans.

- Exercise care in the handling of sharp objects.

- Report all injuries at once to your T.A.

Please honor these rules!

Prep Room

The prep room is under the supervision of Ms. Joanne Chaplin. Students in the course are not allowed inside the prep room because it is a small, crowded facility and staff members working in the room should not be disturbed. Always see your laboratory instructor if you need assistance.

Class Attendance

Materials for any exercise will be available from Monday through Thursday only. Check procedures in the syllabus to make-up a laboratory if you are unable to attend your regular section due to illness or other reasonable excuse.

The lab exercises will usually require the entire $3\frac{1}{2}$ hour period. **Please arrive on time.**

Advice to the Students of Bio 204

I suspect it is unlikely that you will read this paragraph or, if you do, pay much attention to it. But for those that might, I wish to pass along a few words of welcome and advice.

If you become a biology major, Biol. 204 will be a small part of your studies in this department. Nevertheless, organismal biology is an important component of your training and provides a framework for all of your future studies. Now for the advice.

1. Decide now if you want to learn and do well grade-wise. Do not misjudge the 2 hr credit designation. Depending on your background, but for nearly all students, you will need a lot of time for this course.

2. Spend your effort in the following ways:

    (a) Attend all lab lectures. Pay close attention and take good notes. Supplement the lectures by studying the text assignments.

    (b) Do the above and read over the lab manual <u>before</u> your lab meeting.

3. This is really important! As much as you can, <u>do your own work in lab</u>. The temptation will often be to look at a particular specimen on a neighboring microscope, then move on to the next item. It's fine to look at another prep, but find everything yourself. The Midterm exam, will test not only test for memory recognition of items, but for how well you have developed your dissection skills, your ability to make good biological preparations and your skillful use of the microscope to "find things." Although admittedly at an early stage in your biology studies, it is the intent of this course to challenge the full range of your biology knowledge and skills. To meet that challenge, start your preparations with the very first lab.

Welcome to Biology 204. I am looking forward to studying with you.

Jim Riopel and Kristen Curran

I wish to acknowledge Ms. Lara Call (CLAS 1998) and Ms. Linda Johnson (PhD candidate, Biology) for the fine artwork and the editing assistance that they rendered for this edition.

THE MICROSCOPE

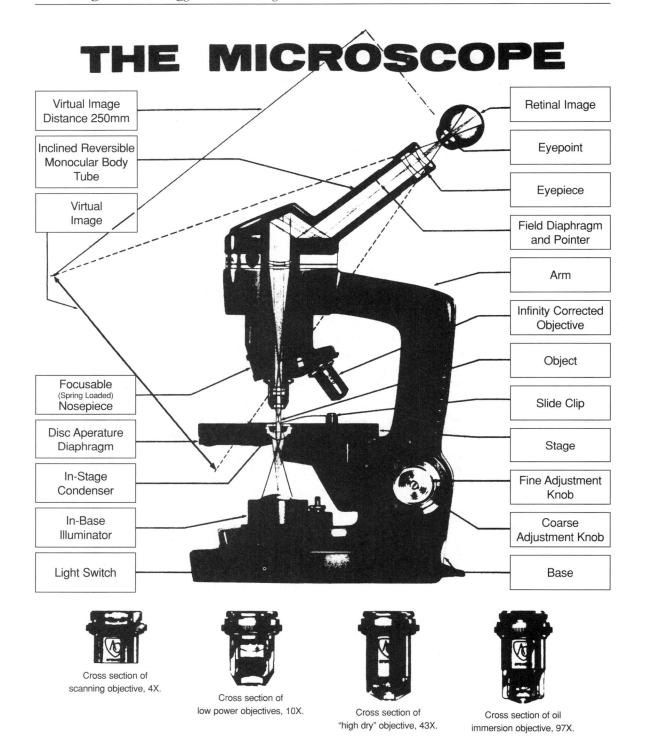

Virtual Image Distance 250mm	Retinal Image
Inclined Reversible Monocular Body Tube	Eyepoint
Virtual Image	Eyepiece
	Field Diaphragm and Pointer
	Arm
	Infinity Corrected Objective
	Object
Focusable (Spring Loaded) Nosepiece	Slide Clip
Disc Aperature Diaphragm	Stage
In-Stage Condenser	Fine Adjustment Knob
In-Base Illuminator	Coarse Adjustment Knob
Light Switch	Base

Cross section of scanning objective, 4X.

Cross section of low power objectives, 10X.

Cross section of "high dry" objective, 43X.

Cross section of oil immersion objective, 97X.

A Typical Compound Microscope

The Protozoa

Objectives:

Satisfactory completion of this unit requires a thorough study of your lecture notes and the laboratory material. The objectives are:

1. Learn the distinguishing characteristics of the phyla and their representatives.

2. Become familiar enough with this group of organisms that you could analyze an unidentified specimen, classify it to major group and describe by discussion and labeled sketches its salient features.

Introduction to Protozoa

As we begin to study the diversity of plants and animals we will begin with less complex forms of life. The protozoa are single-celled organisms. In one cell, all of the functions of life are performed. There are no organs or tissues, but there is a division of labor within the cytoplasm with specific organelles that are specialized to carry out specific tasks.

As we look at protozoa keep in mind specific functions, such as skeletal support, locomotion, sensory perception, reproduction, etc. Consider how these functions are achieved within the domain of one cell.

In your analysis don't lose sight of the importance of these creatures, or their interesting lifestyles. The name protozoa means "first animals" and, indeed, their simple organization suggests that some of the first organisms in evolution were like them. As for their lifestyles, protozoa turn up in some interesting places. In this class there almost certainly will be someone that has a species of amoeba living in their mouth. More conventional (and acceptable!) places are in the soil, or in fresh and salt water.

This exercise introduces some of these interesting creatures. Only a few are presented to provide you with ample time for observation and study.

Use of the Microscope

Obtaining and Caring for the Laboratory Microscope

1. Over the next few weeks you will become very comfortable using the compound microscope. Remember that many students throughout the week will use the microscope that you are using today. For this reason we need you to take responsibility for its care and maintenance.

   a. The compound microscopes and dissecting scopes should always be carried with two hands!

   b. If your light source does not work or there is a problem with your microscope let your TA know before the end of class.

   c. The oculars and objectives at some point need to be cleaned. Sometimes this can easily be done with a dry kimwipe. Always wipe with a clean surface of the kimwipe in one direction only! DO NOT RUB YOUR OCULARS AND OBJECTIVES IN A CIRCULAR MOTION! If the dry kimwipe does not solve your problem, ask your TA for some 70% ethanol. Perform the above steps again with a kimwipe soaked with 70% ethanol and dry in the same manner with a dry kimwipe.

Adjusting the Microscope to Fit You

1. Use a prepared slide of *paramecia* to become familiar with the compound microscope.

2. Ocular Adjustment:

   a. Adjust the distance between the oculars so that a single image under the 10x objective is seen.

   b. Close your left eye and use the right eye to focus on the specimen, then with the right eye closed turn the left ocular adjusting ring for sharp focus with the left eye. The procedure should be done at the beginning of every lab exercise and will reduce headaches!

Light Intensity and Focusing of Light

1. Locate the rheostat to raise or lower the light intensity. This should be regulated during examination of specimens to optimize the viewing. **Note: light intensity can be increased by raising the condenser.**

2. Condenser

   a. Locate the knob for raising and lowering the condenser. Use this while viewing your specimen at 10X and 40X magnification levels. At high levels, the light intensity is increased and focused. This is especially important at 40X. Make sure a grainy light background is avoided. This problem is prevented by slightly raising or lowering the condenser.

b. The iris diaphragm is regulated by a lever on the front of the condenser. This adjustment is very important to optimize critical detail. Close the iris to improve detail. Experiment to see the effect of open and closed diaphragm under 40x while viewing your prepared slide.

Using the Ocular Micrometer to "Size Up" your Microorganism

1. In your right or left ocular is an ocular micrometer. You should see evenly spaced lines across your field of vision.

2. Obtain a stage micrometer (you will need to share!).

3. Place the stage micrometer on the microscope stage and view it through the 4X objective. You need to calibrate all three of your objectives (4X, 10X, and 40X).

4. Focus on the stage micrometer so that you can clearly see the markings on it.

5. Align your ocular micrometer with the stage micrometer (match up the parallel lines).

6. Superimpose the two micrometers using the stage adjustment knobs. Align the left-most marking of the stage micrometer with the zero mark on the ocular micrometer.

7. Count the number of divisions on the stage micrometer that correspond to 10 divisions on the ocular micrometer. Take that number and multiply by the distance between the markings for the stage micrometer (this number is etched on the micrometer itself). Divide that number by 10 to get the actual distance between each division on your ocular micrometer. This is the calibration for the 4X objective!

8. Calibrate the 10X and 40X objectives and use your new found knowledge to analyze the length and width of a paramecium on your prepared slide.

9. Be sure to use this information to estimate the size of organisms you observe today!

Phylum Ciliophora

Background

Ciliophora represent a large group of organisms. There are 3 classes and numerous orders. They are characterized by cilia in at least one stage of their life cycle. They are unique in that they have two types of nuclei: a large macronucleus, essential to survival, and one or more micronuclei that can be removed. Reproduction is by asexual fission or less frequently by sexual conjugation. The majority of ciliates are free-living and occur in many fresh or marine habitats. A few are parasites in humans.

Our first example will be the Paramecium. This protozoan lives in fresh water usually containing much decaying organic matter. They are often found in pond scum.

Our next example, <u>Stentor</u>, is common in freshwater lakes, ponds and streams. Its funnel shape makes it easy to recognize. They are usually sessile, attaching to rocks or detritus. However, if food or other conditions change, the animals can detach and swim free. There are several common species that are quite striking because of their coloration. The most common are pigmented blue; others may be pink, green or gold. Stentors reproduce by binary fission or rarely by conjugation. Experiments have shown that the animal can be cut into several fragments that regenerate provided a piece of the macronucleus is included.

Procedure

Our purpose is to provide an opportunity for you to observe and record as much as you can about <u>Paramecium</u> morphology, responses to stimuli, behavior and reproduction.

1. <u>Morphology</u>. Place a drop of <u>Paramecium caudatum</u> culture on a clean slide. Obtain your sample by pipetting near a wheat seed in the bottom of the jar. Add a coverslip and observe under low power. It will be useful to adjust the light source with the iris diaphragm. Note the general appearance and movements of the animal. It will help for detailed study to slow down the movement. An inert viscous medium can be used to mechanically impede movement of the animal. **Place a drop of methyl cellulose or Protoslo on a slide. Add a drop of water containing <u>Paramecia</u> to the middle of the Protoslo and without mixing, gently add a coverslip.** The <u>Paramecium</u> will slow down as it swims across the interface into the denser medium. With the aid of **Figure 1-1** note the following features in the living specimens:

   a. Observe the morphological outline. The more blunt end is the anterior part.

   b. The <u>oral groove</u> begins at the anterior end and leads to the <u>cytostome</u> or mouth. The <u>cytopharynx</u> is a short, funnel-shaped tube extending posteriorly into the cytoplasm. <u>Food vacuoles</u> develop at the terminal end. The vacuoles are then carried in the cytoplasmic stream (<u>cyclosis</u>) while digestion is achieved. The path of the vacuoles is clockwise, traveling first toward the posterior end, then around to the anterior part. Undigested materials are voided through the anal pore located just posterior to the oral groove. The pore is hard to see except when solid wastes are excreted.

   c. Two clear, pulsating <u>contractile vacuoles</u> can be seen. The radiating arms are difficult to observe. It may help to partially remove water from under the coverslip. Draw water out by placing the edge of a paper towel at the side of the coverslip. The vacuoles are in a fixed position. They regulate water content. It is interesting that freshwater protozoans typically have contractile vacuoles. They are not present in most marine species. Why?

   d. <u>Cilia</u>. Movement is achieved by the rhythmic beating of cilia. Observe cilia at the margin of the animal or in the oral groove. Use high magnification and reduced light.

   e. <u>Pellicle</u>. This is basically the skin of the <u>Paramecium</u>. If you dry out a specimen, a hexagonal pattern can be seen under high magnification.

f. Trichocysts are located at the edge of the ectoplasm. They are spindle-shaped, but hard to see before discharge of a thread-like filament. In some protozoans these structures may serve as defense mechanisms. **Add a drop of 1% tannic acid to the edge of the coverslip. As it reaches the animal, it is possible to observe discharged trichocysts with low illumination and 40X magnification.**

g. Nuclei. Paramecium has a large macronucleus that is involved in metabolic functions such as feeding and digestion and a micronucleus that carries the hereditary genes. The nuclei are difficult to see in living specimens. Prepared slides are available. **Try to stain your own preparation by drawing a drop of acetocarmine under one edge of the coverslip.**

2. Feeding. Paramecia feed on bacteria and other small organic particles. Obtain a sample from the spot plate containing paramecia that have been fed yeast stained with congo red. Apply a coverslip and observe. Note the currents of cilia in the oral groove and the passage of yeast down the groove and into the cytostome. You can see food vacuoles form, reach a certain size, then break away. It may be possible to follow the course of food vacuoles. As digestion proceeds the pH will change. Congo red is red in pH 5.0 or above and turns blue in more acid solutions. **NOTE: this experiment works well, but it takes time. Don't expect instant results. Start preparations early enough so that they can be checked every few minutes or so over a 30–60 minute period.** Before excretion of undigested waste left in the food vacuole, the pH may become more alkaline, returning the color of the vacuole toward a reddish tint.

3. Response to stimuli. Add a few absorbent cotton fibers to the slide. Observe how paramecia react to the barrier.

It is possible to design a variety of experiments to test the response of paramecia to environmental stimuli. **Watch their reaction to a drop of dilute acetic acid added to the edge of the slide.** A similar response can be observed by adding a few grains of salt to the culture.

4. Reproduction. Paramecium divides by binary fission. There is not much to see here. Organelles replicate and the organism divides into two more or less equal units. There are prepared slides available. In Paramecium, the division is transverse. In others it may be longitudinal or variable.

There is also a sexual process called conjugation in Paramecium. Prepared slides of conjugation are also available.

5. Stentor.

a. Observe a jar containing an undisturbed Stentor on the TA desk. If Stentor is undisturbed, they attach and have the characteristic funnel shape. Their shape is otherwise rather amorphous. They are found mostly in bottom debris. (See **Figure 1-4c.**)

b. Sample your own Stentor and note the characteristics of the "Stentor on the move" versus the "undisturbed Stentor."

Paramecium caudatum

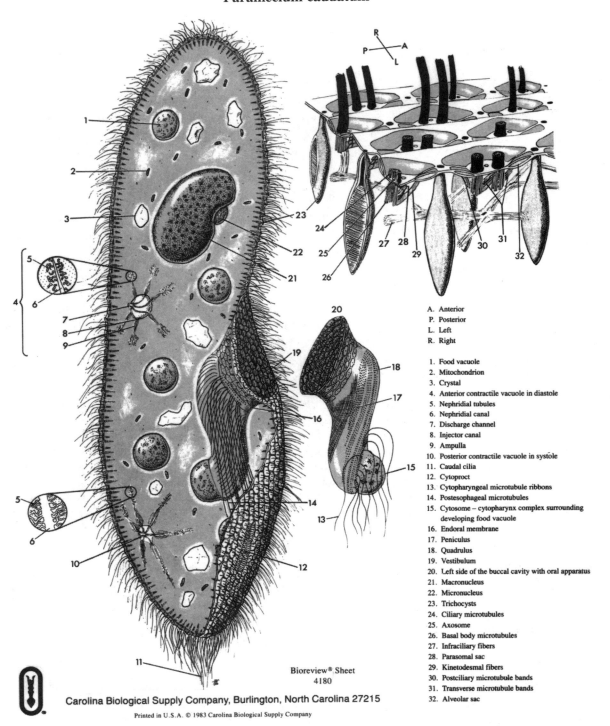

A. Anterior
P. Posterior
L. Left
R. Right

1. Food vacuole
2. Mitochondrion
3. Crystal
4. Anterior contractile vacuole in diastole
5. Nephridial tubules
6. Nephridial canal
7. Discharge channel
8. Injector canal
9. Ampulla
10. Posterior contractile vacuole in systole
11. Caudal cilia
12. Cytoproct
13. Cytopharyngeal microtubule ribbons
14. Postesophageal microtubules
15. Cytosome – cytopharynx complex surrounding developing food vacuole
16. Endoral membrane
17. Peniculus
18. Quadrulus
19. Vestibulum
20. Left side of the buccal cavity with oral apparatus
21. Macronucleus
22. Micronucleus
23. Trichocysts
24. Ciliary microtubules
25. Axosome
26. Basal body microtubules
27. Infraciliary fibers
28. Parasomal sac
29. Kinetodesmal fibers
30. Postciliary microtubule bands
31. Transverse microtubule bands
32. Alveolar sac

Bioreview® Sheet
4180

Carolina Biological Supply Company, Burlington, North Carolina 27215

Printed in U.S.A. © 1983 Carolina Biological Supply Company

Figure 1-1. Paramecium Anatomy

Phylum Rhizopoda

Background

Rhizopod species have pseudopodia for feeding and locomotion. The body may be a naked protoplast or develop an external skeleton. There are both free-living and parasitic species, and the best known are the amoebae. They live in fresh water, salt water, soil, and at Carolina Biological Supply Co. where there they find their way into nearly all introductory high school and college biology courses. If you have seen them before, take time today to look again and follow through on the observations recommended. They are quite extraordinary creatures. They are bottom surface dwellers and must have a substratum on which to move. We will be looking at **Chaos** or the related, but larger **Pelomyxa**, a freshwater species that is usually found in slow moving or still water. They may be on the bottom or on the surface of aquatic plants where they capture and eat algae, bacteria, and other protozoa.

The body of the amoeba is transparent and constantly changes shape as it extends and withdraws pseudopodia. The outer plasmalemma is fringed with fine hair-like projections. Internally, the cytoplasm is differentiated into a thin, clear peripheral layer of ectoplasm just under the cell membrane and a more fluid granular endoplasm. Within the endoplasm there are a variety of inclusions and organelles (see **Figure 1-2**). There is a conversion of ectoplasm to endoplasm and vice versa as pseudopodia are extended. Movement is achieved by the extension of the more liquid plasmasol. At the periphery the plasmasol changes to plasmagel. Apparently the movement in amoebae is similar to muscle contraction. Contractile proteins like the actin and myosin of vertebrate muscles function in the amoeba, polymerizing and depolymerizing in the extension and retraction of pseudopodia.

Reproduction in the animals is asexual by binary fission. The organism simply divides by coordinated karyo- and cytokinesis. Studies of the mitotic process reveal between 500 and 600 chromosomes so that precise details of this process are not known. How old are the specimens under study? This is difficult to say but, grown under optimal conditions, amoebae divide about every 24 hours. They can remain viable under conditions of dormancy for months.

Procedure

Using the dissecting scope on the side bench, **pipette your sample from the bottom of the culture flask where the amoebae are more concentrated.** Place a drop from the amoeba culture to a clean slide. Place a coverslip over the specimen supported by a few grains of sand or pieces of broken coverslip.

Study under low magnification with reduced light to improve detail and to avoid overheating the specimen.

1. <u>Locomotion</u>. Myosin filaments have been found in <u>A. chaos</u>, but regulation of movement is still a mystery. Can you see any evidence of coordinated, directional movement? At one-minute intervals, for 10 minutes, prepare rapid sketches of your animal.

2. <u>Morphology</u>. With the aid of your text and **Figure 1-2** understand the following structures or regions.

   a. <u>Plasmalemma</u>. External cell membrane.

   b. <u>Ectoplasm</u>. Thin layer of clear cytoplasm under the plasmalemma.

   c. <u>Endoplasm</u>. Inner granular region.

   d. <u>Plasmagel</u>. Granular outer layer of endoplasm in the gel state.

   e. <u>Plasmasol</u>. Internal, granular endoplasm in a fluid or sol state. The plasmasol is usually undergoing movement.

   f. <u>Contractile vacuole</u>. A clear vacuole in the endoplasm. Functions in water balance or osmoregulation. Excess water is collected and the vacuole discharges the water through the plasmalemma. A new vacuole then forms. Why would the contractile vacuole have an important function in protozoa (think about what you know from the <u>Paramecium</u>)?

   g. <u>Food vacuoles</u>. Digestion takes place in these structures. They contain enzymes and ingested food particles.

3. <u>Other Sarcodina</u>.

   a. <u>Arcella</u>:

      There are related species of amoebae that have an outer shell or **test**. <u>Arcella</u> occurs in fresh water bays where it feeds on rotting vegetation. The test is made of siliceous or chitinous fragments fixed in place by polymerized proteins (**Figure 1-4a**). They are easily seen under the dissecting scope. They appear as small, brown basketballs. Unfortunately, most of the tests (shells) are empty. The darker ones have someone inside. The tests are abundant and require no special procedure. Simply pipette from the bottom of the dish.

      We also have prepared slides of <u>Foraminifera</u>, another sarcode with a snail-like shell. These species represent the most common group of protozoa in the world. Deposits of these animals are responsible for the so-called white cliffs of Dover, England.

   b. Of course there are many other amoebae. Several species of <u>Entamoeba</u> occur in humans. <u>E. gingivalis</u> is a commensal organism that lives in the mouth along the gum where it feeds on bacteria. We encourage you to check for this. We certainly hope no one has <u>E. histolytica</u>. It lives chiefly in the large intestine where it causes dysentery by feeding on the intestinal lining and red blood cells. The parasite has two phases, the trophozoite, or active feeding stage and the encysted stage. It is passed to a new host in the encysted stage containing 4 nuclei. After ingestion the multinucleated amoeba emerges from the cyst and divides by fission producing 4 uninucleate amoebae. Amoebic dysentery is a world-wide problem, but it is especially so in underdeveloped tropical countries. Farmer (1980) gives a recent account of this parasite.[1]

---

[1]Farmer, J. 1980. <u>The Protozoa</u>. Introduction to Protozoology. C.V. Mosby Co.

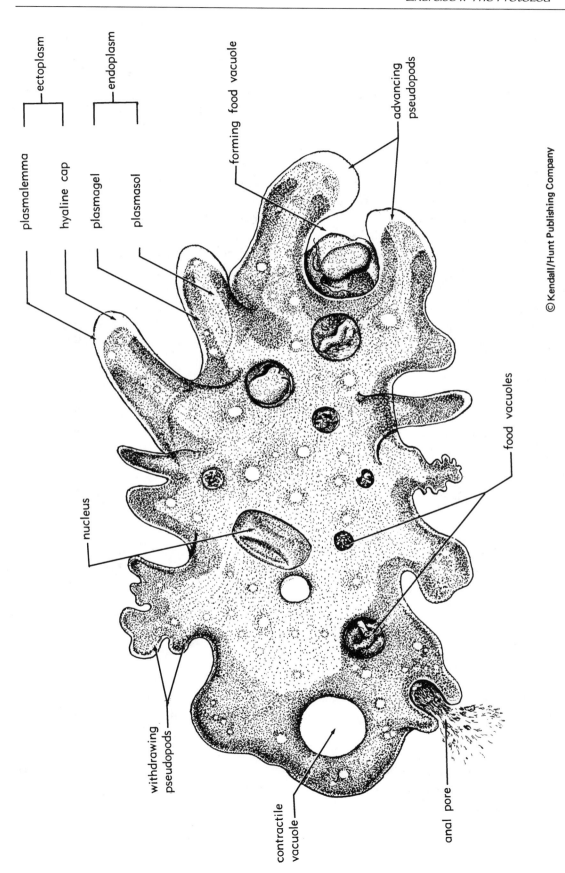

ectoplasm

endoplasm

plasmalemma

hyaline cap

plasmagel

plasmasol

forming food vacuole

advancing pseudopods

nucleus

food vacuoles

withdrawing pseudopods

contractile vacuole

anal pore

Figure 1-2. Amoeba

E. histolytica is almost always present where sanitary conditions are not maintained, yet it is capable of cropping up in populations where sanitary conditions are considered to be of a high order. One breakdown in sanitary conditions at the right time and at the right place can aid in its dissemination. The classical example of this in the United States is the well-known case of the outbreak of amoebic dysentery in 1933 during the Century of Progress Exposition in Chicago in which thousands of tourists visited the city, thus increasing the chances of contact between carriers and hosts. The beginning of the outbreak was traced to a hotel where a faulty plumbing connection permitted an ice-water tank to become contaminated with sewage. More than a thousand cases of amoebic dysentery resulted, and 58 known deaths were attributed to infections of E. histolytica. A more recent example is an epidemic that started among employees of a factory in South Bend, Indiana. Approximately half of a work force of 1,500 individuals was affected with several deaths being attributed to E. histolytica. Apparently a drinking water line became contaminated with sewage at a leaking point in the pipe. With the increased facility and speed of travel and the pressure for tourism in our modern world, it is difficult to imagine a decrease in the incidence of this parasite. In fact, repetition of the 1933 epidemic with a much broader distribution is possible. In the case of the Chicago incident, the parasite was known to be transported from Chicago to forty-four states and three Canadian provinces. If a similar outbreak were to occur at the present time, the parasite could have worldwide distribution within 24 hours, with hosts being jetted out of O'Hare International Airport to all parts of the world.

Phylum Mastigophora (Gr. *Mastix*, "Whip")

Background

Species of this phyla use flagella for locomotion. There are both free-living and symbiotic species. Reproduction is usually asexual by fission.

Example: **Trichonympha**. A symbiont of termites and wood roaches. This is a good example of symbiotic mutualism. Neither organism could live without the other. Trichonympha can digest the cellulose of wood eaten by termites.

Termites not infected with this flagellate will continue to eat wood, but they cannot digest it and die. It is interesting that there are no cysts of Trichonympha carried through the moult of the termite. The mode for reinfection is not certain, but always takes place in the presence of other infected termites. If newly hatched larvae are isolated they never acquire any protozoa. It is apparently different with wood-eating roaches (Cryptocercus spp.) which have a permanent population of about 25 species of flagellates, many of which form cysts at moulting stages.

We also have the opportunity to observe another small flagellate, Chilomonas (**Figure 1-4b**). This common freshwater species has two flagella, usually slightly unequal in length. It makes its living from decaying vegetative matter, but is in turn a food source for other heterotrophs.

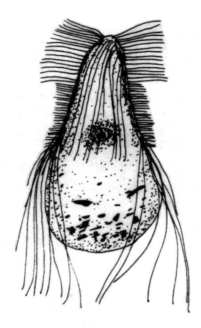

Figure 1-3. Trichonympha

Procedure

1. Before trying a live preparation of <u>Trichonympha</u> it would be wise to view a prepared slide of species of flagellates found in the termite gut and see if you can identify a <u>Trichonympha</u>! What objectives do you need to use?

2. Live preparation:

    Place a termite on a slide and use a razor blade or needle to decapitate. Next, cut off the posterior two-thirds of the abdomen and place in a drop of 0.6% saline solution (NaCl). Use your scalpel to make a small cut in the posterior of the animal then use a glass pipette or needle to roll out the contents of the abdomen onto the slide. Place a coverslip on the specimen and gently depress the coverslip to flatten out the prep. Add more saline if needed to fill out the space under the coverslip.

    Examine your prep. Make sure your illumination is adjusted satisfactorily. Use a partially closed iris diaphragm. Practice locating <u>Trichonympha</u> with the 4X and 10X objectives. Note other species not to be confused with <u>Trichonympha</u>. Consider mode of locomotion and details of structure and size. Record your observations.

    (Before proceeding take a moment to reflect on how this course will work. During the midterm exam you will be asked to prepare specimens and show the TA your preparation (such as Trichonympha). Have you prepared sufficiently to do that later? It is important for you to grasp early on the full extent of how to conduct these studies.)

3. Look for <u>Chilomonas</u> in your sample of <u>Paramecium caudatum</u>, most likely being chased or engulfed! Although it is truly without pigment, <u>Chilomonas</u> may appear slightly blue due to the numerous, conspicuous starch granules that fill most of the endoplasm. The nucleus may be visible due to its relatively large size.

Phylum Actinopoda

Background

Members of this group are primarily freshwater counterparts of the marine Radiolaria. Both have unique pseudopodia called axopodia that function in feeding. Bacteria, plankton and other microorganisms adhere to the sticky surface of the axopodia and are quickly moved along to the cell body region where they are engulfed into the cytoplasm—not a pleasant outcome to contemplate if you are a microorganism! In the case of Actinosphaerium, larger creatures like copepods entangled by several axopodia are then engulfed by pseudopodia. In some cases several individuals have been observed to participate in the capture of larger prey.

Procedure

Actinosphaerium

This is a beautiful fresh water species (see **Figure 1-4a**) and requires **careful** pipetting. Locate under the dissecting scope at high magnification. Actinosphaerium appear as small grayish spheres, and are usually found with debris on the bottom of the dish. Use the curved pipettes to transfer to a slide.

Phylum Bacillariophyta

Background

These organisms are also known as diatoms. They are unicellular and are found in both fresh and salt water. Diatoms are usually yellow or brown in color. The color of a diatom comes from the photosynthetic pigments within the cell. The presence of these pigments allows this organism to use the energy from the sun to make its own food. Diatoms are themselves an important source of food to many marine and fresh water creatures. It is estimated that 20–25% of all organic carbon fixation on the planet is carried out by diatoms. These organisms deposit silica in their cell wall forming a shell called a frustule. Why do you think diatoms have chosen to form a shell? Do they have cilia or flagella? Do you think they may have any control over their movements in the water? Why don't they just sink?

Procedure

1. Centrate and Pennate diatoms:

    Diatoms are found in two basic shapes. Centrate diatoms are generally radially symmetrical while pennate diatoms are bilaterally symmetrical. Note the differences between these morphologies as you sample and observe these organisms.

2. Diatomaceous earth (diatomite):

    The richest source of diatom fossils is found in deposits of their skeletons in diatomite, or diatomaceous earth. These are the skeletons of ancient diatoms that died and settled at the bottom of lakes and oceans. Sample some diatomaceous earth and note the variety of diatom skeletons found there.

Review Questions

1. <u>Streblomastix strix</u> (illustrated at right) is a protozoan some of you may recognize because it inhabits the termite gut.

    a. Based on external morphology, what phylum is it in?

    b. What kingdom?

2. Where can <u>Entamoeba histolytica</u> be found?

3.a. Some amoebae carry a "shell" around with them. What is the shell structure called?

    b. Give the genus of an example we studied that had this shell.

4. What is the approximate size of an amoeba?

5. Name two functions of pseudopodia.

6. Connect:

    <u>Entamoeba gingivalis</u>   mutualism

    <u>Paramecium</u>   causes bad breath

    <u>Foraminifera</u>   found on the ocean floor

    <u>Trichonympha</u>   macro and micronucleus

Answers

1.a. Mastigophora

   b. Protista

2.   Human intestine

3.a. Test

   b. <u>Arcella</u> or <u>Difflugia</u>

4.   200–600 mm

5.a. Locomotion

   b. Feeding

6.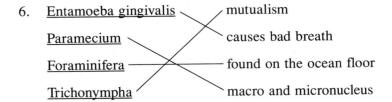

References

Borradaile, L.A. and F.A. Potts. *The Invertebrata*. Cambridge University Press, London, 1961.

Brown, F.A., Jr. (ed). *Selected Invertebrate Types*. John Wiley and Sons, New York, 1950.

Corliss, J. *The Ciliated Protozoa*. Pergamon Press, 1961.

Farmer, J. *The Protozoa*. C.V. Mosby, 1980.

Hyman, L.H. *The Invertebrates,* Vol, I. Protozoa Through Ctenophora. McGraw Hill, New York, 1940.

Jahn, T.L. and F.F. Jahn. *How to Know the Protozoa*. W.C. Brown, Co., Dubuque, Iowa, 1949.

Pennak, R.W. *Fresh Water Invertebrates of the United States*. Ronald Press, 1953.

Purves, et al. *Life: The Science of Biology,* 6th edition. Sinauer Associates, Inc., Sunderland, Massachusetts, 2001.

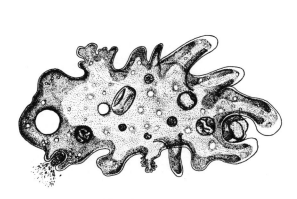

Figure A. Amoeba

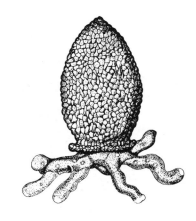

Figure B. Difflugia

Lateral View

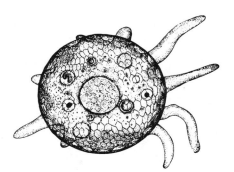

Surface View

Figure C. Arcella

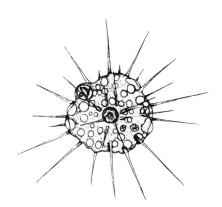

Figure D. Actinophrys

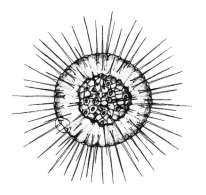

Figure E. Actinosphaerium

(Not drawn to scale)

PROTOZOA, SARCODINA

Figure 1-4a. Representative Protozoa

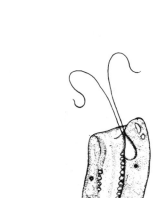

Figure A. Chilomonas

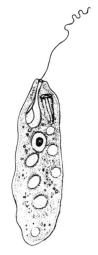

Figure B. Peranema

Figure C. Euglena

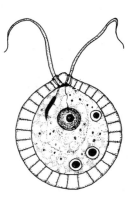

Figure D. Hematococcus

Figure E. Phacus

(Not drawn to scale)

PROTOZOA, FLAGELLATE

Figure 1-4b. Representative Protozoa

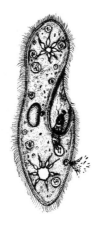

Figure A. Paramecium

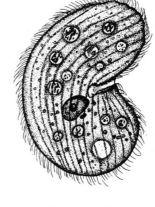

Figure B. Colpoda

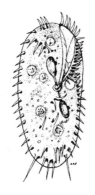

Figure C. Stylonichia

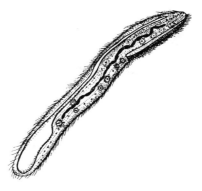

Figure D. Spirostomum

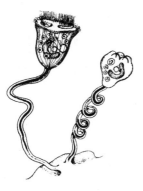

Figure E. Vorticella

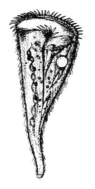

Figure F. Stentor

(Not drawn to scale)

PROTOZOA, CILIATE

Figure 1-4c. Representative Protozoa

Bio 204 Review Checklist

Lab 1—Kingdom __PROTISTA__

Phylum MASTIGOPHORA	*Trichonympha*	flagella	symbiotic*
	Chilomonas	biflagellate	starch granules
Phylum RHIZOPODA (aka SARCODINA)	*Chaos,* or *Pelomyxa*	pseudopods endoplasm plasmogel plasmalemma phagocytosis	fission ectoplasm plasmosol contractile vacuole
	Arcella	test	
	Foraminifera	test	
Phylum CILIOPHORA	*Paramecium*	cilia cytostome cyclosis pellicle macronucleus fission	oral groove cytopharynx contractile vacuole trichocysts micronucleus conjugation
	Stentor	blue-ish	
Phylum ACTINOPODA	*Actinosphaerium*	axopodia	
Phylum BACILLARIOPHYTA	diatoms	photosynthetic pennate plankton	silica in cell wall centrate

---

*Items marked with an asterisk are not items that can be identified, per se, but are terminology (processes, characteristics, etc.) with which you should be familiar!

The Fungi

Exercise

2

Objectives:

Satisfactory knowledge of this unit requires thorough study of the text and completion of the laboratory exercises. The objectives are to:

1. Provide "hands-on" experience in microscopic observation and record-keeping by comments, sketches and drawings.

2. Learn the salient features that distinguish fungi from other organisms.

3. Know the names and distinguishing characteristics of the divisions of fungi.

4. Learn the generalized life cycle of each representative fungus.

5. Become prepared to analyze an unknown fungus, classify it to major group and describe its salient biological features.

Introduction

Fungi are the principle organisms along with bacteria that recycle organic matter into reusable forms. This decomposition process is of vital importance to a balanced ecosystem in which CO_2 is released and nitrogen and minerals are recycled. These enormous benefits are somewhat balanced by the non-selective diet of fungi. Almost anything organic can be decomposed by these organisms including wood, paper, leather, clothes, even crude oil.

Fungi differ from plants in that they are unable to photosynthesize. They are either <u>saprophytes</u>, which live on dead organic matter in the soil, or <u>parasites</u>, which obtain their food directly from other living organisms. Some fungi parasitize animals (causal organisms of "athlete's foot" and other types of dermatitis), but many more attack plants. For example, Dutch elm disease, oak wilt, potato blight, and corn and wheat rusts were all caused by parasitic fungi.

Fungi are <u>heterotrophic</u> organisms. Most fungi consist of masses of filaments called <u>hyphae</u> (collectively termed <u>mycelium</u>). The fungi resemble plants because of the presence of cell walls. Reproduction in the fungi is by spores (motile or nonmotile), which may be produced sexually or asexually. Fungal reproduction may also occur by fission, budding, and fragmentation. Fungal reproductive structures are not covered by a layer of protective cells.

In all fungi, the mycelium is the vegetative part of the organism, since it carries on all the general activities of cells, such as absorption, digestion, respiration, and excretion. However, it does not carry on photosynthesis, since the fungi have no chlorophyll. Because the fungi literally live on and in their food supply, and because they must make this food soluble so it can be taken into their protoplasm, they have a capacity to produce a vast number of enzymes, many of which are unlike those of higher plants. This adaptation enables them to break down many complex substances for their use such as cellulose which cannot be used by most other organisms. The by-products of enzyme digestion are then absorbed into fungal cells.

Reproduction in fungi is usually cued to one or more environmental factors. Nutrient availability, temperature and moisture are important signals. It is the sexual and asexual reproductive structures which, as a rule, constitute the part of the fungus most commonly seen. For example, the mushroom, as you know it, is the spore-bearing structure. The widespread network of mycelia is underground and seldom seen. The rust, which you may have seen on the leaves of wheat, represents chiefly the spore-bearing hyphae. The vast and intricate mycelia penetrates deep into the tissues of the plant. It is the mycelium that damages the plant, but from a standpoint of fungus control, it is the tremendous number of spores that constitutes the potential danger for the thousands of plants not yet infected.

As you study the various groups note the trend away from multinucleate mycelia and toward more complex reproductive structures. Note similarities and differences between these taxa.

Summary of General Fungal Characteristics

1. Non-photosynthetic. Heterotrophic. Obtain nourishment from other food. Extra-cellular digestion. Enzyme degradation of substrate followed by absorption.

2. Cell wall of chitin. Many but not all species have cell walls composed of the polysaccharide chitin which is the same material as in the exoskeleton of insects and crustaceans.

3. Filamentous body form and tip growth. No specialized tissues, but fungi develop hyphae which may aggregate into species-specific reproductive structures that are unique in form and color.

Classification of Fungi

Members of the Kingdom Fungi are classified into five principal divisions: Chytridiomycota, Oomycota, Zygomycota, Ascomycota and Basiodiomycota. Separation is based on characteristics of the hyphae and features of reproduction (see Table 2-1). An additional taxon Deuteromycota or Fungi Imperfecti, includes fungi that do not have an observable sexual stage. Chytridiomycota and Oomycota are now considered to be Protists.

Table 2-1. The Kingdom Fungi

DIVISION	NUMBER OF SPECIES	EXAMPLES	DISTINCTIVE CHARACTERISTICS	DISEASES	POSITIVE ECONOMICAL USES
Chytridiomycota	About 650	Chytridium, Allomyces	Most aquatic; non-septate hyphae; flagellated spores and gametes; cell walls contain chitin; vegetative body usually a thallus, rarely hyphal	Lettuce big vein; viruslike infection of corn	None
Oomycota	About 475	Potato blight fungus, Saprolegnia	Non-septate hyphae. Some aquatic; flagellated spores; formation of eggs and sperm in special gametangial; cell walls contain cellulose. Gametangial meiosis; gametes only stage that is haploid	Blights and mildews of plants; fish infections	None
Zygomycota	About 600	Black bread mold Rhizopus	Formation of zygospores (tough, few, resistant spores resulting in a fusion of gametangia); no flagellated cells	Few	None
Ascomycota	30,000	Neurospora, yeasts, morels, truffles, Sordaria	Formation of fine asexual spores (conidia); sexual spores in asci; hyphae divided by perforated septa; dikaryons; no flagellated cells	Powdery mildews of fruits, chestnut blight, Dutch elm disease, ergot	Food morels, truffles, wine, beer, bread-making (yeasts)
Basidiomycota	25,000	Toadstools, mushrooms, rusts, smuts, Coprinus	Sexual spores in basidia; hyphae divided by perforated septa; dikaryons; no flagellated cells	Rusts, smuts	Food (mushroom)
Deuteromycota	25,000	Penicillium	Fungi with no known sexual cycles; no flagellated cells	Thrush	Cheeses, antibodies

Chitridiomycota (the chytrids)

Background

Chytrids are microscopic with a simple morphology; the body of most species consists of a single cell, sometimes bearing hypha-like branches. In asexual reproduction, the contents of a cell divide into a group of zoospores, which in some species exhibit amoeboid movements, and in others move by ciliary motion. In certain species, the zoospores exhibit a combination of ciliary and amoeboid movement. Sexual reproduction has also been reported in chytrids, but it is apparently much less frequent than asexual reproduction in most species. Most chytrids are parasites of algae, microscopic animals, insects, higher fungi, and seed plants. The body of the fungus may be attached to the surface of the host, or it may inhabit a single host cell. Some of the parasitic chytrids cause serious diseases of higher plants; for example, the potato wart diseases (caused by Synchitrium endobioticum), a leaf and fruit disease of cranberries, and the brown-spot disease of corn.

Allomyces is an interesting microorganism and is an example of this group. In nature this fungus grows in soil on plant and animal remains. If you collect soil samples and put the soil in petri plates with dead seeds, Allomyces will often attack them. You can almost be certain if not Allomyces some other chytrid or water mold will be present in the soil sample.

Allomyces is ideal for our study. It is stationary, but has motile reproductive stages. Although multicellular, its body is composed of branching filaments that are one cell in thickness. This makes it easy to observe cytological features of growth, differentiation, and sexual and asexual reproduction. Note also that the life cycle is like higher plants in expressing an alternating haploid, gamete-producing stage and a diploid, spore-producing stage. What are the advantages of this alternation?

In this study you will observe both stages and learn how to trigger the switch from vegetative growth to reproductive growth. This will be an easy lab manipulation. Think of how it relates to the natural environment of Allomyces.

From the diploid culture you will see asexual zoospores released from sporangia and swimming away, later to establish a new fungus. The haploid fungus will develop gametangia and discharge motile gametes. We expect that most students will see the fusion stage of the sex cells.

Procedures have been worked out so that you will have several cultures at defined developmental stages making it possible to visually review most of the life stages of Allomyces in this lab period. Study **Figure 2-1** before proceeding.

Procedure

One to two weeks old 1n and 2n agar cultures of Allomyces will be available. With a scalpel or toothpick, scrape off a small amount from one culture of Allomyces and place it in water on a one half of a slide. Cover with a coverslip. Place a sample of the other culture on the other half

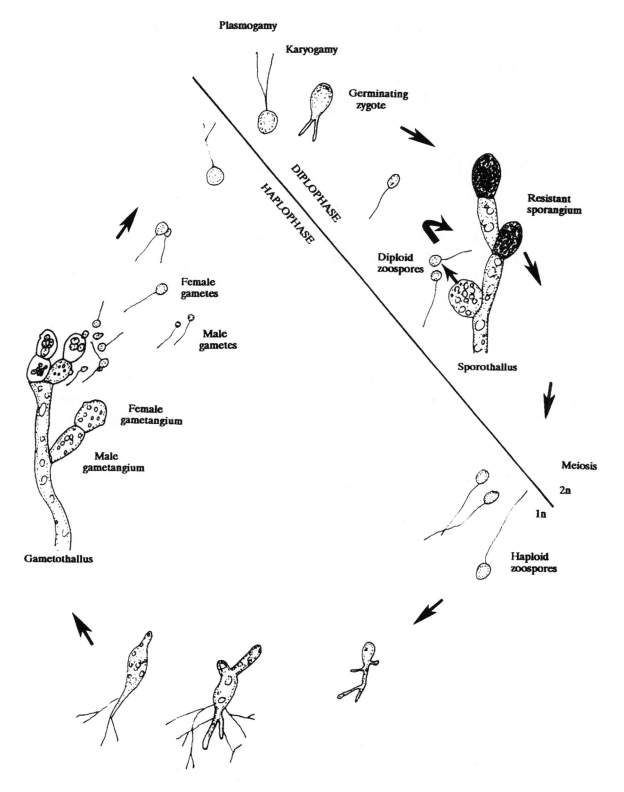

Figure 2-1. Allomyces Life Cycle

of the slide in the same manner. Label each half of the slide with a permanent marker and store in a moist chamber (petri dish with moist filter paper) for about one hour, then examine. You should see gamete release and movement in the haploid phase. Careful observation may also show fusion stages of gametes. Note consistent size, positional, and color differences of male and female gametangia. In the diploid phase, observe resistant sporangia and motile zoospores.

Division Oomycota (Gr. oion, meaning egg)

Background

Water molds are found in almost all bodies of water where they live mostly on dead plant and animal remains. Oomycetes ("egg fungi") get their name from their sexual cycle in which large non-motile eggs are produced inside a special structure called an <u>oogonium</u> (**Figures 2-2 and 2-3**). Egg fungi are also called by several common names, including water molds, algaelike fungi, or downy mildews.

Unlike other fungi, whose cell walls are composed of chitin, the cell walls of oomycetes are made up of cellulose.

If you have ever had an aquarium, you are probably more familiar with the oomycetes than you realize. These molds often attack diseased or dying fish. One oomycete, <u>Phytophthora</u>, was responsible for the potato famine in Ireland in the 1850's. Another, <u>Plasmophara</u>, almost destroyed the French wine industry.

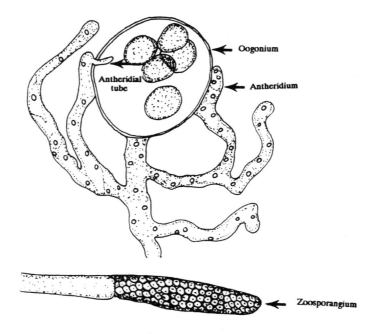

Figure 2-2. <u>Saprolegnia</u> Reproduction

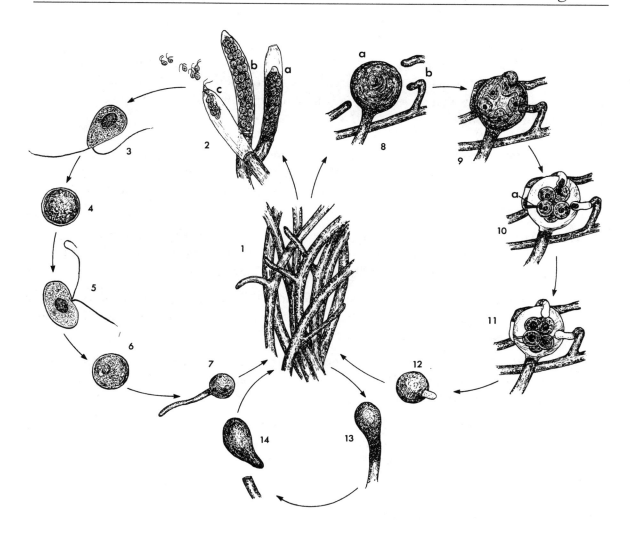

1. Somatic hyphae
2. Sporangia
   a. Proliferation
   b. Mature sporangium
   c. Discharge of primary zoospores
3. Primary zoospore
4. Aplanospore
5. Secondary zoospore
6. Aplanaspore
7. Germination
8. Developing sexual structures
   a. Oogonium
   b. Antheridium
9. Cleavage of oospheres
10. Fertilization
    a. Fertilization tube
11. Oospores
12. Germination
13. Gemma
14. Germinating gemma

Carolina Biological Supply Company, Burlington, North Carolina 27215

Printed in U.S.A.© 1967 Carolina Biological Supply Company

Bioreview® Sheet
8332

Figure 2-3. <u>Saprolegnia</u> Life Cycle

Procedure

We will be observing <u>Saprolegnia</u>. This is a water mold quite common in pond and creek water. It is usually saprophytic but may attack fish and aquatic insects. Be sure to identify the structural characteristics that are typical of Oomycota and distinguish between asexual and sexual reproductive structures.

1. To orient yourself, first locate the oogonia and antheridia in a prepared slide. Look carefully for the fertilization tube that extends from the antheridium to one of the egg cells. Also try to find an oogonium in which there are several thick-walled zygotes (oospheres).

2. Observe the development of live mycelium on seeds that have earlier been placed in pond water. Some of the tips of hyphae may appear whiter and denser than other portions of the mycelium. Microscopic examination will reveal the reason for this.

3. With a pair of forceps, remove a small amount of the mycelium. Mount in water under a cover glass. Study the hyphae carefully at 10X and 40X. Look for cross walls at the base of organs, zoosporangia and zoospores.

4. Living material may show motile zoospores. When released the spores are pear-shaped with two terminal flagella. They swim for some time and then settle down and the flagella are withdrawn and a resistant stage occurs.

   Prepare drawings of: a portion of the mycelium
                                zoosporangium
                                sexual reproduction stage

Division Zygomycota (Bread molds)

Background

Bread molds are filamentous, like most fungi, and have long, tube-like cells called hyphae (singular, hypha) which, taken in the aggregate, form the fuzzy mycelium. The vegetative hyphae in this division have no cross walls and are multinucleate.

We will be observing a typical Zygomycota, <u>Rhizopus stolonifer</u>, the black bread mold. It reproduces both sexually and asexually. The latter process results in the formation of a sporangium, a stalked, saclike structure containing the asexual sporangiospores. In addition to maturing sporangia, you may also be able to find sporangia that have ruptured and released their sporangiospores, leaving a swollen columella behind. In sexual reproduction, the hyphae of two different mating types must come together. Specialized side branches arise from the point of contact of the two hyphae and adhere to each other at their tips. Cross walls are laid down enclosing several nuclei in each tip, after which the walls between the two tips break down and their contents fuse. The resulting cell, called a zygospore, enlarges and develops a thick, dark-colored wall.

Procedure

1. Identify the structural characteristics typical of zygomycetes, and distinguish between asexual and sexual reproductive structures.

2. With the aid of **Figure 2-4**, study sexual and asexual structures on prepared slides.

3. Use your dissecting scope to compare growth of live cultures of <u>Rhizopus</u>. Make a wet mount and use your light microscope to observe as much cellular detail (illustrated in **Figure 2-4**) as possible.

Division Ascomycota (Ascos = sac)

Background

<u>Ascomycetes</u>, or sac fungi, (along with the basidiomycetes) are often referred to as "higher fungi" because their hyphae are made up of uninucleate cells partitioned by cell walls ("septa"). Septate walls are perforate allowing some nuclear migration. Hyphae of Zygomycota and Oomycota are coenocytic (multinucleate) without crosswalls.

Ascomycetes include some familiar fungi such as morels, truffles, <u>Sordaria</u>, <u>Neurospora</u>, and yeasts. <u>Claviceps</u> produces a mycelium causing ergot on rye grass. Perhaps one of the most well-known ergot derived products is LSD. Of course yeasts produce ethyl alcohol by the process of fermentation and thus are important in the production of beer and wine. Chestnut blight, Dutch elm disease, apple scab, brown rot in fruit, and powdery mildew infections are serious plant diseases caused by the ascomycetes.

Asexual reproduction in the ascomycetes occurs by the mitotic production of haploid <u>conidiospores</u> which are produced in long chains at the end of specialized hyphae called <u>conidiophores</u>.

Sexual reproduction in the ascomycetes results in the development of an ascus containing four to eight ascospores (four are produced by meiosis; additional spores are the result of mitosis). A group of asci are usually found together in a fruiting body known as an ascocarp (**Figure 2-4**). The ascocarp occurs in a variety of shapes by which different ascomycetes can be identified.

Procedure

1. For this Division you need to be able to identify the structures typical of ascomycetes, distinguish between mechanisms of asexual and sexual reproduction and explain the basis for the name, "Ascomycota."

    *Refer to your text for the life cycle of the ascomycetes.

2. Use a dissecting microscope to examine living material from <u>Sordaria</u>, and observe prepared slides at 40X. Sketch and label the fruiting bodies (ascocarps), asci, and ascospores of the ascomycete, <u>Sordaria</u> (**Figure 2-6**).

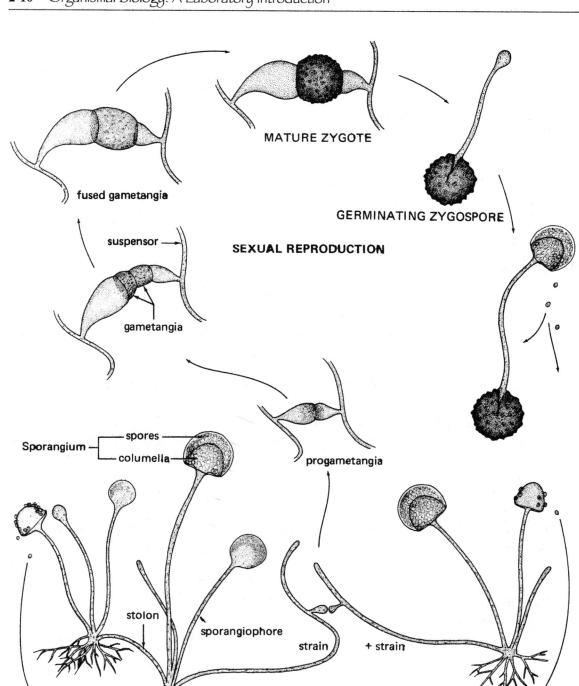

MATURE ZYGOTE

fused gametangia

GERMINATING ZYGOSPORE

suspensor

SEXUAL REPRODUCTION

gametangia

Sporangium — spores
columella

progametangia

stolon

sporangiophore

strain   + strain

rhizoids

ASEXUAL REPRODUCTION

Figure 2-4. Life Cycle of <u>Rhizopus</u>

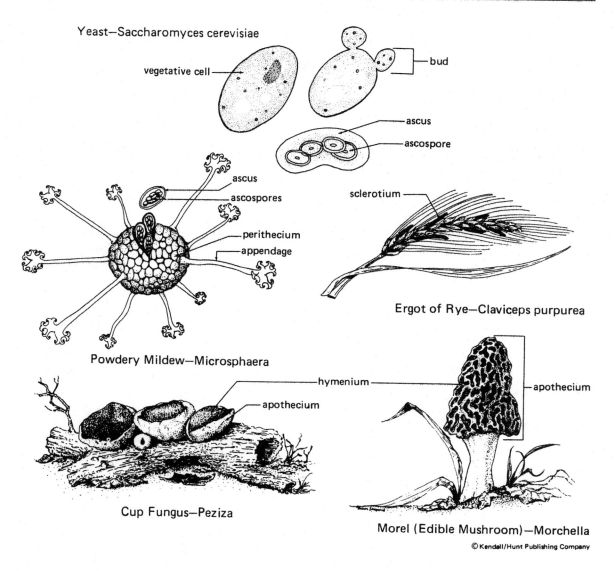

Figure 2-5. Types of Ascomycetes

3. Remove several of the small, black, round fruiting bodies from a culture of Sordaria and prepare a wet-mount slide. Press on the coverslip with the eraser end of your pencil until the ascocarps pop. This will release the asci. Study the slide at high power (40X). Draw and label your observations.

    a. How many ascospores are present in each ascus?

    b. By what division process (or processes) were the ascospores produced?

4. Examine prepared slides of yeast. Notice the buds on some of the unicellular organisms. Instead of a hyphal filament, yeasts are unicellular and can reproduce asexually by budding or sexually by the production of asci (**Figure 2-5**).

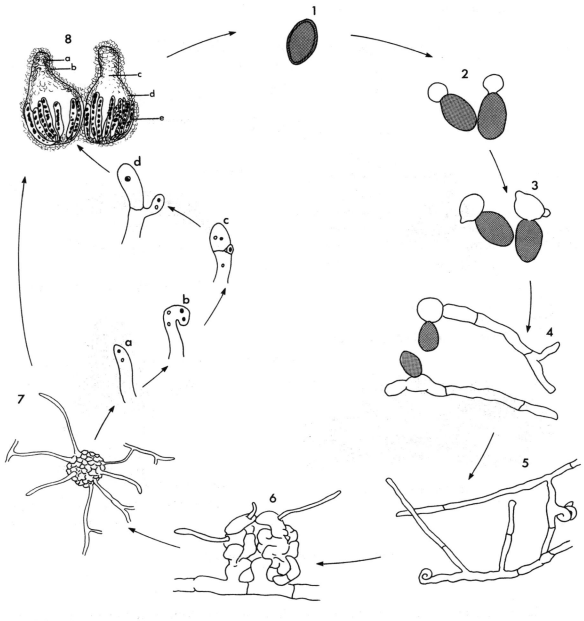

1. Ascospore
2. Ascospore germination
3. Development of hyphae
4. Proliferation and branching of hyphae
5. Development of crosier
6. Early protoperithecial development
7. Protoperithecium
   a. - d. ascus development

8. Mature perithecium, l. s.
   a. periphyses
   b. neck
   c. neck canal
   d. perithecial wall
   e. ascus

Carolina Biological Supply Company, Burlington, North Carolina 27215

Printed in U.S.A. © 1974 Carolina Biological Supply Company

Figure 2-6. Sordaria Life Cycle

Division Basidiomycota (Basidion = small base)

Background

Most of the fleshy fungi we know are found in this group. The <u>Basidiomycetes</u>, or club fungi, have a separate mycelium like the ascomycetes, but they differ from the ascomycetes in having sexual spores (basidiospores) borne externally on a club-shaped structure, the basidium, instead of within a sac.

The fruiting body of the basidiomycetes is the basidiocarp. In the higher basidiomycota, the mature basidiocarp may develop a large number of pores or gills on its underside. These gills contain numerous clubshaped basidia, single cells that produce basidiospores. Basidiocarps come in many different sizes, shapes, and colors. Some, such as the mushrooms with which you are familiar, are edible.

Rusts, smuts, puffballs, toadstools, and shelf fungi are also members of the division Basidiomycota. Many species form mycorrhizal associations with plants, in which case a symbiotic relationship develops between fungal hyphae and plant roots, providing both the plants and fungi with important nutritional elements.

Procedure

1. Be able to identify the structures typical of basidiomycetes and understand how a "mushroom" is formed.

2. Obtain a fresh edible mushroom, <u>Portobello</u>, and examine it carefully. Identify the parts described below (see **Figures 2-7, 2-8**).

   <u>Cap</u>—The umbrellashaped portion of the fruiting body (basidiocarp).

   <u>Gills</u>—Radiating strips of tissue (<u>lamellae</u>) on the undersurface of the cap; basidia form on the surface of the gills.

   <u>Basidia</u>—Clubshaped, sporeproducing structures on the surface of the gills.

   <u>Basidiospore</u>—A spore produced by meiosis on the outside of a basidium.

   <u>Stalk</u>—The upright portion of the fruiting body that supports the cap—a mycelium composed of many intertwined hyphae.

   <u>Ring</u> (or <u>annulus</u>)—A membrane surrounding the stalk of the fruiting body at the point where the unexpanded cap was attached to the stalk.

3. Remove a small portion of the cap with several gills attached. While holding them together, cut a very thin crosssection through the gills with a razor blade (when you are finished it should look like false eyelashes). You should slice from the cap into the gills, not the reverse! Put sections in a drop of water on a slide and apply a coverslip. You

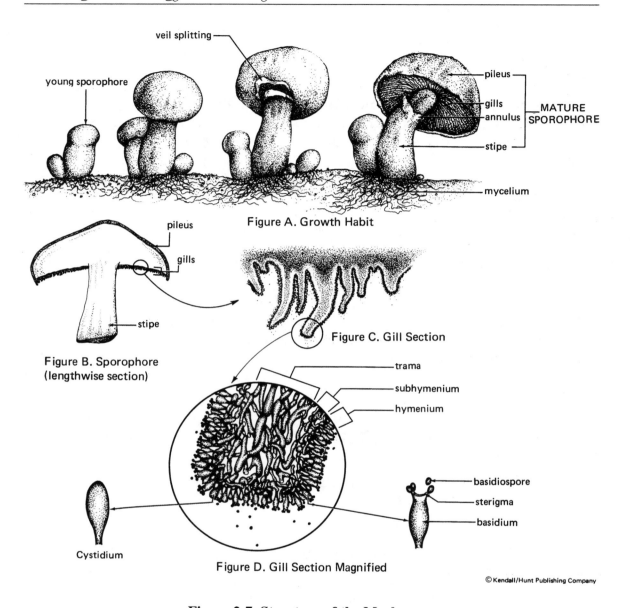

Figure A. Growth Habit

Figure B. Sporophore
(lengthwise section)

Figure C. Gill Section

Figure D. Gill Section Magnified

© Kendall/Hunt Publishing Company

Figure 2-7. Structure of the Mushroom

should be able to see basidia along the edges of the gills. Study your slide at high power (40X). Your TA will demonstrate this procedure for you.

a. Can you see the hyphae making up the thickness of the gills?

b. Are they septate?

c. Are the basidia clubshaped?

4. For a clearer view of the reproductive structures of basidiomycetes, examine prepared slides of a cross-section through the cap of the basidiomycete <u>Coprinus</u>. Find basidia and basidiospores.

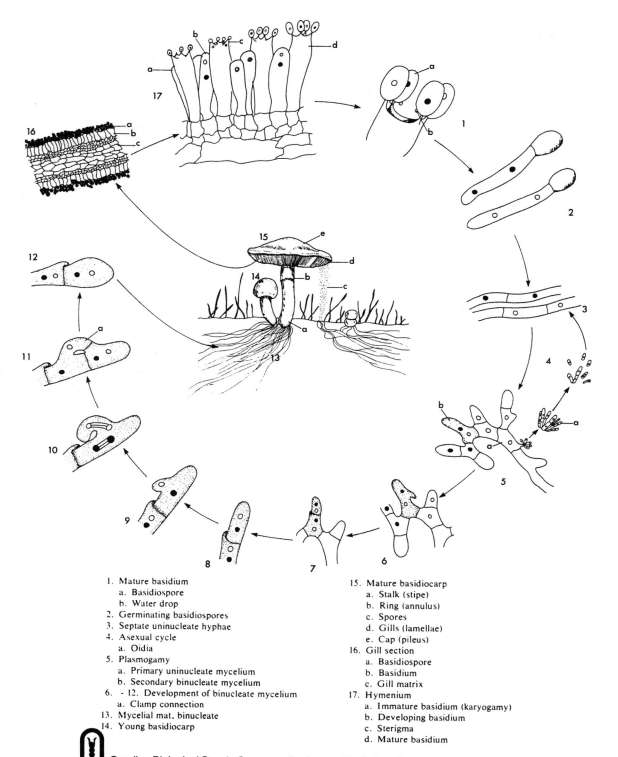

1. Mature basidium
   a. Basidiospore
   b. Water drop
2. Germinating basidiospores
3. Septate uninucleate hyphae
4. Asexual cycle
   a. Oidia
5. Plasmogamy
   a. Primary uninucleate mycelium
   b. Secondary binucleate mycelium
6. - 12. Development of binucleate mycelium
   a. Clamp connection
13. Mycelial mat, binucleate
14. Young basidiocarp

15. Mature basidiocarp
    a. Stalk (stipe)
    b. Ring (annulus)
    c. Spores
    d. Gills (lamellae)
    e. Cap (pileus)
16. Gill section
    a. Basidiospore
    b. Basidium
    c. Gill matrix
17. Hymenium
    a. Immature basidium (karyogamy)
    b. Developing basidium
    c. Sterigma
    d. Mature basidium

Carolina Biological Supply Company, Burlington, North Carolina 27215
Printed in U.S.A. © 1975 Carolina Biological Supply Company

Figure 2-8. Gill Fungus Life Cycle

Division Deuteromycota

Background

Deuteromycetes, the "imperfect" fungi, are those fungi in which the sexual stages are not known to exist. This may be because these fungi have not been completely studied, or because the sexual stages have truly been lost during the course of evolution. Reproduction is asexual by conidia (**Figure 2-9**) which are produced at the tips or sides of haploid hyphae rather than within sporangia. Examples in this taxon include blue molds and green molds, some of which are important sources of antibiotics. Trichophyton causes athlete's foot. We will look two different forms of Deuteromycetes, a yeast form and a hyphal form. The first is Rhodotorula that represents a yeast form. It is easily recognized by its beautiful coral color. The other is Penicillium, which, as the name implies, is the source of the antibiotic penicillin.

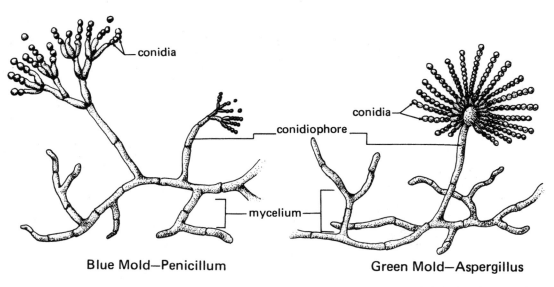

Blue Mold—Penicillum   Green Mold—Aspergillus

Figure 2-9. Examples of Conidia

Procedure

Penicillium

1. Obtain a clean slide.

2. Add two drops of Lactophenol Cotton Blue dye to your clean slide.

3. Obtain a piece of scotch tape 1–2 inches long. Hold the tape between your fingers in the in the shape of a "U" with the sticky side out.

4. Open the lid of the Penicillium plate and gently touch the tape to the growing fungus. Close the lid.

5. Lay the tape fungus side down on the slide so that the fungus is in the dye (the tape acts as your coverslip!)

6. Blot off any dye that squeezes our from under the tape and view the morphology of the fungus.

7. Prepared slides of conidiophores of <u>Penicillium</u> and <u>Aspergillus</u> are available. **The <u>Aspergillus</u> slides are unusually <u>thick</u>, so care MUST be taken to avoid damaging the slides or the objectives!!**

Rhodotorula

1. Prepare 2 samples of <u>Rhodotorula</u> on one slide.

   a. Prepare the first sample as a normal wet mount.

   b. Prepare the other sample as a wet mount, but add 1 drop of Methylene blue dye before coverslipping.

2. Which preparation gives you the most information about the morphology of <u>Rhodotorula</u>?

Diversity Among the Lichens

Background

   <u>Lichens</u> are distinct organisms that are actually two organisms in one. The body is made up of alga cells (usually a green or blue-green alga) embedded in the mycelium of a fungus (usually an ascomycete or basidiomycete). The fungus is the dominant (most prominent) of the two organisms. Thus, lichens are usually studied with the kingdom Fungi.

   Lichens are found on tree trunks, rocks, and arctic mountaintops, to name just a few locations. Often lichens are the first colonists on bare, rocky areas. Be sure to understand the nature of the symbiotic relationship of the lichen organism.

Procedure

1. Note the growth forms of lichens on demonstration. See if you can identify the following three types: <u>crustose</u>, closely encrusting bodies; <u>foliose</u>, leafy bodies; <u>fruticose</u>, shrubby, branching bodies.

2. Examine a prepared slide on demo of a lichen thallus showing algal cells surrounded by fungal hyphae (see **Figure 2-10**).

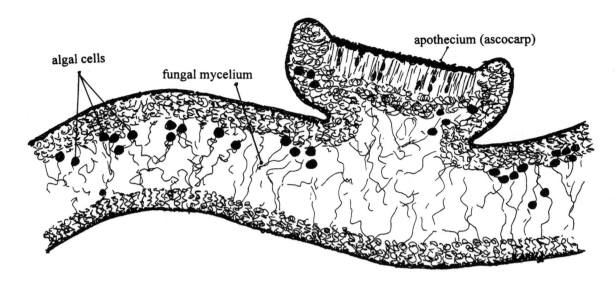

Figure 2-10. Lichen Thallus

Review Questions

1. Fungi have cell walls composed principally of _____ .

2. Fungi are ecologically important as _____ .

3. Name a member of the Deuteromycota.

4. Where do the sexual spores develop in <u>Sordaria</u>? In <u>Coprinus</u> or <u>Agaricus</u>?

5. A zygospore is the end product of the fusion of _____ .

6. _____ are the only fungi with flagellated male and female gametes.

7. Draw the fruiting body of a Basidiomycota.

8. Lichens are symbiotic associations of _____ and _____ .

Answer Section

1. <u>Chitin</u>

2. <u>Decomposers</u>

3. <u>Pencillium</u>; <u>Aspergillus</u>

4. <u>Sordaria—ascus. Coprinus or Agaricus—basidium</u>

5. <u>+ and – gametangia</u>; <u>or just gametangia</u>

6. <u>Chytridiomycota</u>

7. <u>See lab manual pg. 15 or text</u>

8. <u>A fungus usually an ascomycete, and a green alga or a Cyanobacterium</u>

Additional Reading

Alexopoulos C.J. and C.W. Mims. 1979. *Introductory Mycology.* Wiley, New York. (The standard text for many years in Mycology).

Betra, L.R. (Ed.) 1979. *Insect-Fungus Symbiosis: Nutrition, Mutualism and Commensalism.* John Wiley and Sons, Inc., New York. A fascinating collection of articles about the many and varied interactions between these abundant and successful organisms.

Burnett, J.H. 1968. *Fundamentals of Mycology.* St. Martins, New York (p. 22, 546 pp., illus.).

Christensen, C.M. 1975. *Molds, Mushrooms and Mycotoxins.* University of Minnesota Press, Minneapolis. An engagingly written account of some fungi and their importance to humans; supplements, rather than duplicates, his earlier book.

Courtney, Booth, and H.H. Burdsall, Jr. 1984. *A Field Guide to Mushrooms and their Relatives.* Van Nostrand Reinhold, New York. A beautifully illustrated guide to some 350 common species of mushrooms of temperate North America.

Demain, A.L. 1981. "Industrial Microbiology." *Science 214:* 987–995. Valuable review of the prospects for greatly expanded use of fungi and bacteria in industrial processes.

Emerson, R. 1954. *The Biology of Water Molds.* Chap. 8 (pp. 171–208) in *Aspects of Synthesis and Order in Growth.* Ed. by D. Rudnick, Soc. Study Develop and Growth, 13th Symposium. Princeton University Press.

Hale, M.E. 1983. *The Biology of Lichens,* 3rd ed. University Park Press, Baltimore, MD. An outstanding, concise summary of all aspects of the morphology, physiology, systematics, ecology, and economic uses and applications of the lichens.

Koch, W.J. 1973. *Plants in the Laboratory.* MacMillan Co. Good book for further experiments with fungi and a list of references to 1970.

Bio 204 Review Checklist

Lab 2—Kingdom FUNGI   *indicates non-ID terminology

**Know all the life cycles and the sexual/asexual reproductive phases!

General:		heterotrophic*	saprophytic*
		cross wall/septum	zoospores
		hyphae	mycelium
		uni-/multinucleate	gametes
		chitin*	fruiting body
		decomposers*	
Division CHYTRIDIOMYCOTA	*Allomyces*	male and female gametangium (1n)	
		resistant sporangium (2n)	
		gametes	
Division OOMYCOTA	*Saprolegnia*	antheridium	oogonium
		zoosporangium	oospores
Division ZYGOMYCOTA	*Rhizopus*	zygote/zygospore	sporangium
		columella	gametangia
Division ASCOMYCOTA	*Sordaria*	ascus	ascospore
	yeasts	septate	multinucleate
		conidiospores	ascocarp
Division BASIDIOMYCOTA	*Coprinus,* or *Agaricus*	basidium	basidiospore
		sterigma (-ta)	cystidium
		cap (pileus)	gills (lamellae)
		stalk (stipe)	ring (annulus)
		septate	
Division DEUTEROMYCOTA (aka FUNGI IMPERFECTI)	*Aspergillus* *Penicillium* *Rhodotorula*	conidia	only asexual*
		no sporangium*	
		yeast form	
LICHEN		algal cells	mycelium
		crustose	fruticose
		foliose	thallus

---

*Items marked with an asterisk are not items that can be identified, per se, but are terminology (processes, characteristics, etc.) with which you should be familiar.

Evolution of Plants

Exercise
3

Objectives:

The purpose of this exercise is to explore the evolution of plants from aquatic forms (green algae) to terrestrial. We begin with the green algae and then examine the diversity of vascular plants now represented by relatives of ancestral lines by plants from five divisions: <u>Bryophyta</u>, <u>Psilophyta</u>, <u>Lycophyta</u>, <u>Sphenophyta</u>, and <u>Pterophyta</u>. We will then learn about the evolution of the seed by observing the life cycles of Coniferophyta and Anthophyta.

Introduction

Green Algae: The Ancestors of Plants

The algae are actually members of the Protist Kingdom. We are studying them here because the plants we know today evolved from green algae (Chlorophyta). These are extraordinarily beautiful organisms. Their ecological significance is underscored. They are major primary food producers in marine, shoreline and inland freshwater systems. Indeed about half of all photosynthetic organic material is produced by algae.

Algae evolved more than 450 million years ago. Textbook treatments separate algae into a number of taxonomic groups based on pigments, cell characteristics and reproduction. It is likely that algae represent several unrelated groups that possess many of the same features.

Time constraints limit our study to a few representatives. Refer to **Figure 3-1** for illustrations of the species under study.

Kingdom Plantae

Five hundred million years ago (mya) our oceans teemed with life. Invertebrates of all sizes, algae, fungi and bacteria were all present. Sunlight was abundant, but there were no land plants to absorb it. Evidence from our fossils show that the move to land occurred over the next 100 million years, and by the Silurian Period (438–408 mya) vascular plants were in North America and in other parts of the world. By the Devonian (408–360 mya) vascular plants became numerous and diverse.

Land plants probably evolved from an aquatic algae species. How was this achieved? Protection from desiccation, efficient internal water transport and improved reproductive systems all were important requirements. Key events were:

1. The development of spores with their protective walls. Spores can withstand severe temperature shifts, fungal invasion and most important, they protect the internal protoplasm from water loss.

2. Increased body size with the synthetic ability of plant cells to make cutin to coat the outside of epidermal cells. Concomitantly, the need for gaseous exchange through this impervious layer for photosynthesis was solved by the formation of stomates.

3. The development of specialized elongate cells for the conduction of water and sugars, the xylem and phloem.

4. Protection of the gamete producing organs from the environment.

5. The protection of the embryo from the environment. This was the seed, a new organ that also aided in dissemination.

We can only speculate exactly when and how all of these events took place. Indeed we are still guessing on some pieces of the puzzle.

Nonvascular Plants

The first land plants we will investigate are the Bryophytes. These plants lack a vascular system. So they must absorb water from their immediate surroundings. Because they have no vascular system to carry water and food they do not grow very tall and tend to live in moist, marshy environments. They also display a key adaptation to life on land in that their life cycle displays alternation of generations. This means that plants alternate between two multicellular forms one that is diploid and one that is haploid. This occurs in all plants.

The phylum or division Bryophyta contains small, autotrophic plants that have adapted to life on land. Characteristics include:

1. They are green and therefore contain chlorophyll.

2. They have rhizoids, which are root-like structures.

3. They lack vascular tissue

4. The plant body (thallus) may be flattened. In liverworts we see a bilaterally symmetrical thallus while in the mosses the thallus is radially symmetrical.

5. They require water for fertilization to occur and their reproductive organs are multicellular. Archegonium is the female structure and the antheridium is the male structure.

6. They have alternation of generations with the haploid (or gametophyte) stage predomi-nating over the sporophyte (or diploid) stage. The gametophyte can live autonomously while the sporophyte is dependent on the gametophyte for food and water.

Vascular Plants

These plants have vascular tissue and they are common in various parts of the world. Their ancestry can be traced back in most cases for millions of years. Vascular plants are characterized by having a dominant sporophyte generation. The sporophyte is differentiated into true roots, stems, and leaves, each with their own types of specialized tissues. Vascular plants are divided between those that produce seeds and those that do not. The seedless vascular plants are more primitive then those that can produce seeds and include plants such as ferns. These primitive vascular plants still depend on water for sexual reproduction to occur.

Plants with Seeds

In ferns, spores are shed and an independent gametophyte is formed. This biologically frag-ile stage has become reduced during evolution to only a few cells in the seed plants. Fertilization and early growth of the next sporophyte receives protection and continued nourishment from the mother plant. The young sporophyte is the embryo. It is housed in a new structure the seed.

Adaptation: Carnivorous Plants

Many of the interesting features we see in plants are due to their adapting (evolving) to deal with a particular environment. Carnivorous plants commonly live in environments where the soil is depleted of nitrogen. They capture, kill, and digest invertebrates (and the occasional verte-brate). Hundreds of species of plants from several genera are carnivorous. To finish our exercise we will observe some different strategies used by carnivorous plants to supplement their nitrogen intake.

Phylum Chlorophyta

Background

There are more than 7000 known species of this very diverse group. Most are aquatic, but it is hard to find a habitat without them. For example, a species of Chlamydomonas lives on the sur-face of snow. There are others present on tree bark, rocks, soil and some that live as symbionts with lichens, protozoa and invertebrates.

Green algae are considered an important piece in the puzzle of evolution. Several character-istics of the greens are shared with higher plants. Cells of both groups contain chlorophylls a and b, store starch inside plastids, and have very similar wall chemistry. Flagellated gametes of higher plants and algae are also similar.

Procedure

Examine representative species, noting their appearance and size. You may need to be able to recognize these examples and find them again. Prepare wet mounts. As with the protozoa, it is usually most effective to pipette from the bottom of the sample tube.

1. Volvox

    a. Here is a spectacular species. This is a colonial form. Volvox consists of a hollow sphere of biflagellated cells ranging in number from hundreds to 50–60,000. Most are vegetative cells with a few cells specialized for reproduction. Reproduction can be asexual or sexual. Volvox is oogamous (large, non-motile egg and smaller, motile sperm).

    b. Prepare a wet mount with a few grains of sand to elevate the coverslip. Note that flagella movement is coordinated so the entire colony spins clockwise while moving forward. The colony also has polarity with anterior and posterior ends.

2. Spirogyra

    This is a multicellular, filamentous form. It is in common freshwater pond scum. It is rather slippery to touch because the cells are covered with a mucilaginous sheath. Spirogyra is also common to introductory biology labs so you may have seen it before. Note the drawing which shows structural details: the spiral chloroplast, prominent pyrenoids, and the nucleus. Reproduction is asexual or sexual by fusion of conjugation tubes. The variety of form in algae is further illustrated by the last two species. They are frequent visitors to freshwater ponds.

3. Scenedesmus

    This is a colony of 4–8 cells arranged in a flat plate. Each cell contains a nucleus and a large chloroplast with a prominent pyrenoid. Reproduction is asexual.

4. Micrasterias

    This is one example of a desmid. There are thousands of species with many common to our freshwater ponds. They are characterized by a cell deeply constricted along its median giving two semi-cells containing a chloroplast and a connecting zone (the isthmus) containing the nucleus. Reproduction is asexual or by conjugation similar to Spirogyra.

Bryophyta: Liverworts, Hornworts, and Mosses

Background

What we will see in terrestrial plants is the evolution toward being able to survive out of the water. The first Phylum or division we will study is Bryophyta and includes the liverworts, hornworts, and mosses. We will concentrate on an example of a liverwort and an example of a moss. One of the key adaptations to living on land was to create a place for fertilization to occur on

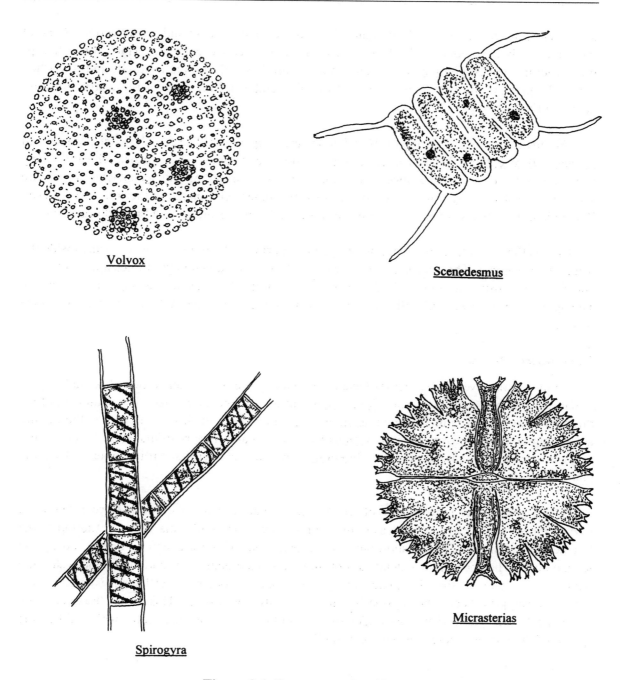

Volvox

Scenedesmus

Spirogyra

Micrasterias

Figure 3-1. Representative Algae

land (<u>embryophyte condition</u>). We will study the life cycles of these particular plants to better understand how they adapted to a terrestrial existence.

Class Hepaticae (Liverworts)

Most liverworts live in moist environments generally in a shaded area. We will be observing *Marchantia,* which displays a flat, green, dichotomously branched thallus that is anchored to the substrate by rhizoids. Marchantia has both asexual and sexual reproduction. Asexual reproduction is accomplished by forming asexual bodies called gemmae, which form in cup-like structures called gemmae cups. The gemmae are basically miniature thalli and if washed out of the cup by rain will form a new thallus. Remember all the cells of these structures are haploid (1N)!

The male and female forms of *Marchantia* arise separately. The female plant produces an archegonial receptacle that has finger like projections at its apex. The egg or archegonia hang from the undersides of these fingers. The male plants form antheridial receptacles, which are stalks with disks on top. On the top of the disk the inner cells of the antheridia form sperm. Remember, no meiosis is needed here because the cells are arising from a haploid organism.

Rain is the catalyst for the meeting of egg and sperm. A diploid sporophyte forms inside the original archegonia. The sporophyte extends a foot into the gametophyte to get food and water and forms a capsule around itself. Meiosis occurs within the capsule yielding haploid spores. The spores are released and half the spores form female plants and half the spores form male plants.

Class Musci (Mosses)

Similar to the liverworts, mosses have a dominant gametophyte generation. The thallus germinates from spores to first form a protonema that is thin and threadlike. The protonema then thickens to form a structure resembling stems and leaves and rhizoids, thus yielding the mature gametophyte. Some mosses are monoecious (one house) and have both male and female structures in the same plant. Others are dioecious (two houses) with separate male and female plants.

At the apex of the male plant antheridia develop that are similar to the liverwort, but stand free of the surface. Sperm are released into rain or dew. The archegonia forms at the top of the female plant. The developing sporophyte then grows out of the archegonia and is fed by the female plant. The sporophyte extends a foot into the female plant to provide food and water and begins to elongate. At first the archegonia grows to accommodate it, but later it is torn off and forms the calyptra. When the capsule is mature the calyptra falls off. This brings the operculum and capsule into plain view. Just under the operculum are peristome teeth, which aid in the spread of the haploid spores during a dry spell.

Procedure (Liverworts)

1. Use a dissecting scope examine fresh and preserved specimens of Marchantia gametophyte thallus, gemmae cups, archegonial receptacles and mature sporophyte. Refer to figure. No sampling please.

2. Examine prepared slides of the *Marchantia:* antheridial receptacles, archegonial receptacles.

3. Learn the life cycle of the *Marchantia* (**Figure 3-2**)!

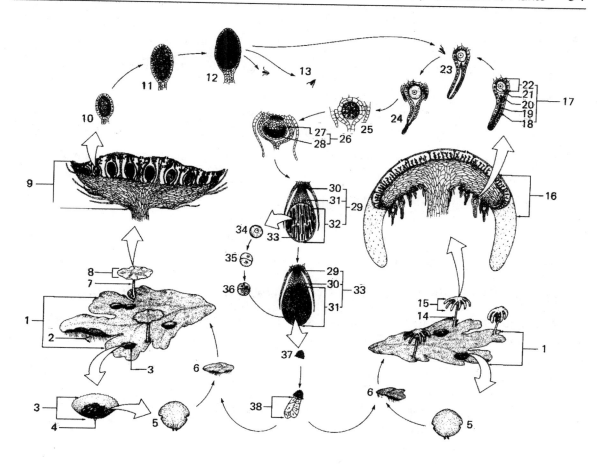

1. Mature Gametophyte Thallus	16. Archegonial Head (L.S.)	29. SPOROPHYTE MATURING
2. Rhizoids	17. Archegonium (L.S.)	30. FOOT
3. Cupule	18. Neck Cells	31. STALK
4. Gemmae	19. Neck Canal Cells	32. CAPSULE
3-6. Asexual Reproduction	20. Ventral Canal Cells	33. ELATER
5. Gemma Released from Cupule	21. Egg Cell	34. SPORE MOTHER CELL
6. Young Gametophyte Thallus	22. Venter	34-36. *Meiosis*
7. Antheridiophore	23. *Syngamy within Archegonium*	35. Dyad
8. Antheridial Head	24. Archegonium with ZYGOTE	36. Tetrad of Meiospores
9. Antheridial Head (L.S.)	25. Archegonium with EARLY SPOROPHYTE	37. Meiospore Released
10. Young Antheridium	26. YOUNG SPOROPHYTE: EARLY TISSUE DIFFERENTIATION	38. Young Gametophyte Thallus
11. Maturing Antheridium	27. STERILE TISSUE	
12. Mature Antheridium	28. SPOROGENEOUS TISSUE	
13. Sperm		
14. Archegoniophore		
15. Archegonial Head		

Figure 3-2. Life Cycle of Marchantia

Procedure (Mosses)

1. Look at the prepared slides of moss archegonial venter and attached stalk, antheridial head, and protonema. Find these structures.

2. With a dissecting scope examine fresh and preserved moss gametophytes with sporophytes. Note the sporophyte's stalk, capsule, calyptra, operculum, and peristome teeth.

3. Each row of student's should break open a dried capsule, share the spores, and view the spores under the microscope.

4. Learn the life cycle of a moss (**Figure 3-3**).

Seedless Vascular Plants

Procedure (Divisions Psilophyta, Lycophyta, Sphenophyta, Pterophyta)

Examine living specimens of these plants. Note distinguishing features. **Figures 3-4 and 3-5** will be useful for details of the club mosses and Equisetum.

1. Psilophyta
Psilotum:
    a. General appearance; absence of leaves and roots.
    b. Location of sporangia
    c. Homosporous

2. Lycophyta
Selaginella:
    a. Note creeping stem habit and spiral arrangement of microphyllous leaves. Identify rhizome (horizontal underground stem) and the roots.
    b. Presence of strobili. Selaginella is heterosporous. Study longitudinal slides of Selaginella. Contrast the size of microspores and megaspores. Understand how development of the gametophytes and the life cycle proceed.

3. Sphenophyta
Equisetum:
    a. Under the dissecting scope, note the ribbed stem and whorled leaves at each node. All species of the Sphenophyta have these characteristics.

4. **Prepared slides:** Compare cross-sections of *Psilotum* and *Equisetum* stem with the *Mnium* stem (a moss or bryophyte).

Background (DIVISION PTEROPHYTA)

Ferns are the most primitive plants that have extensive vascularization in both the stem and the leaves. The leaves are often large and the branching systems of vascular bundles are easily seen. Many fern leaves show a structure in which a subdivision will look like the whole leaf. This repeti-

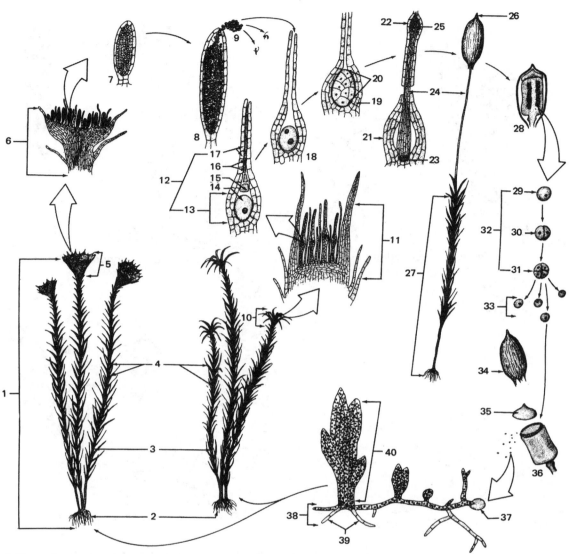

1. Mature Gametophyte Plant
   2. Rhizoids
   3. Non-Vascular "Stem"
   4. Non-Vascular "Leaves"
   5. Antheridial Head
6. Antheridial Head (L.S.)
7. Maturing Antheridium
8. Mature Antheridium
9. Sperm
10. Archegonial Head
11. Archegonial Head (L.S.)
12. Mature Archegonium
    13. Venter
    14. Egg Cell
    15. Ventral Canal Cell

16. Neck Canal Cells
17. Neck Cells
18. ZYGOTE Formed in Syngamy
19. Archegonium Venter Wall
20. EARLY SPOROPHYTE
21. Enlarged Venter Wall
22. Neck Cells Developing into Calyptra
23-25. YOUNG SPOROPHYTE
    23. FOOT
    24. STALK
    25. CAPSULE
26. Calyptra Covering CAPSULE
27. Female Gametophyte with Attached
    Maturing SPOROPHYTE

28. CAPSULE WITH SPORE MOTHER CELLS
29. SPORE MOTHER CELL
30. Dyad
31. Tetrad of Meiospores
32. Meiosis
33. Meiospore
34. Calyptra Removed from Capsule
35. Capsule Lid
36. Meiospores Released from Capsule
37. Germinated Meiospore
38 Protonema
39. Rhizoids
40. Leafy Shoot

Figure 3-3. Life Cycle of a Moss

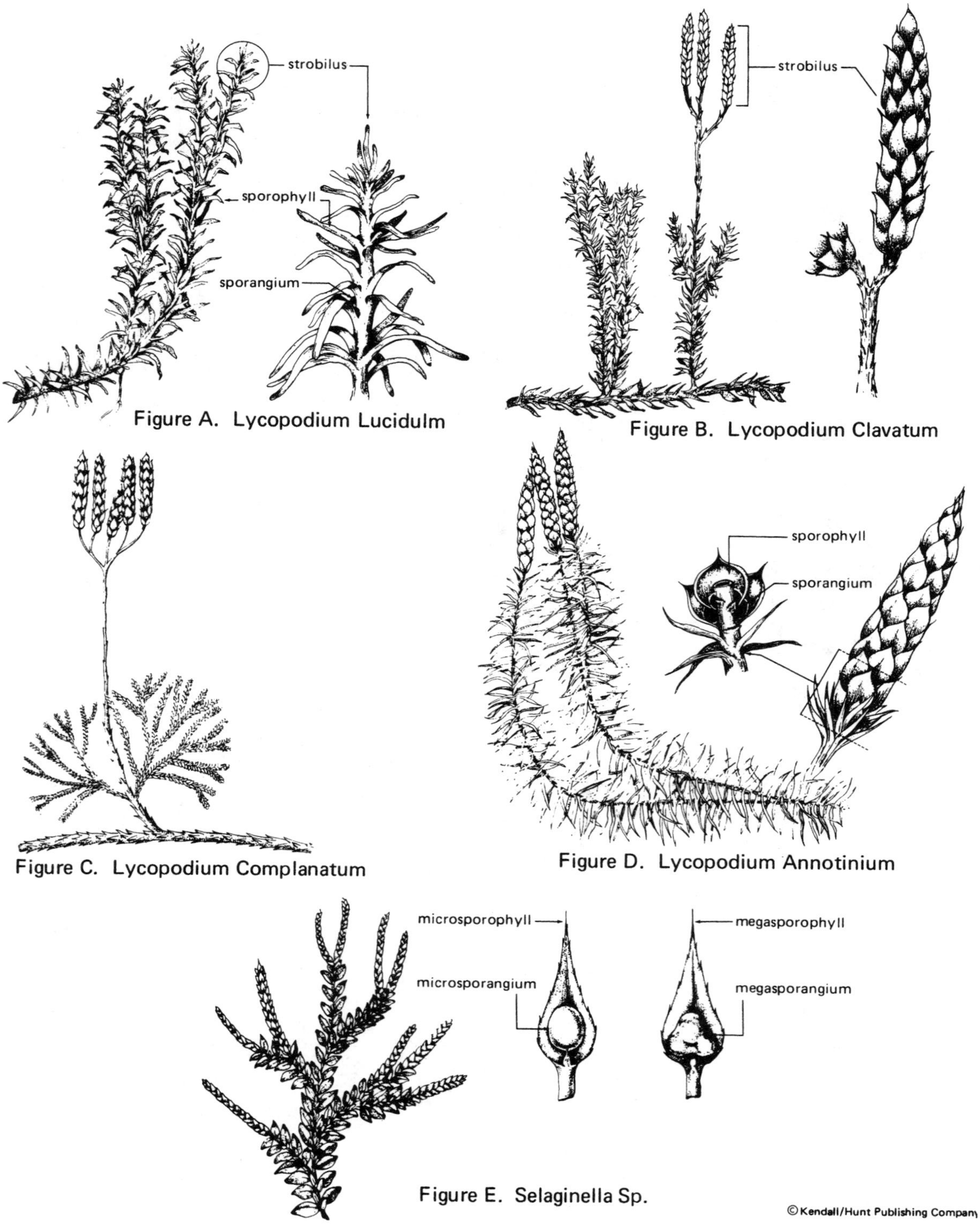

strobilus

sporophyll

sporangium

Figure A. Lycopodium Lucidulm

strobilus

Figure B. Lycopodium Clavatum

Figure C. Lycopodium Complanatum

sporophyll

sporangium

Figure D. Lycopodium Annotinium

microsporophyll

microsporangium

megasporophyll

megasporangium

Figure E. Selaginella Sp.

Figure 3-4. Types of Club Mosses

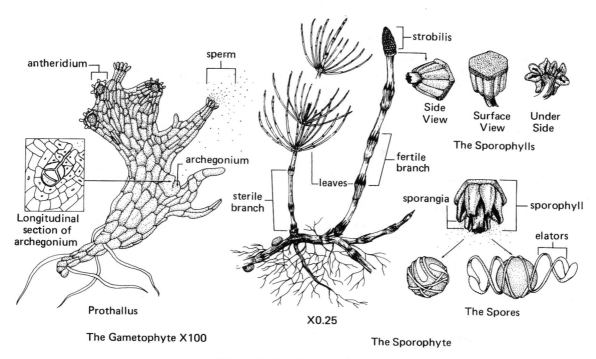

Figure A. Equisetum arvense

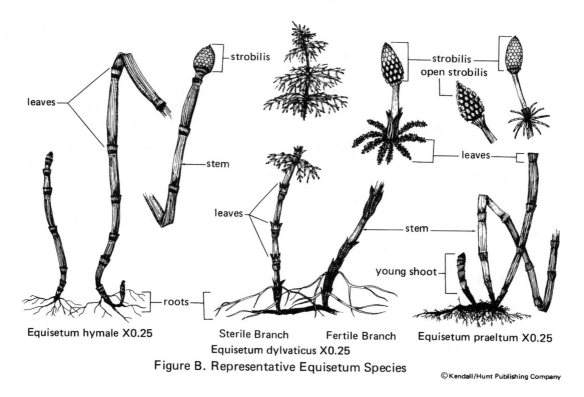

Figure B. Representative Equisetum Species

Figure 3-5. Equisetum: Vegetative and Reproductive

tion continues down to the very small tips of the leaf. The ferns comprise a series of plants with considerable diversity of growth and habitat, such as the large leaf tree ferns with erect stems that grow in tropical rain forests. As a group, the ferns love moisture and shade although a few inhabit fissures in rocks or other dry locations located in bright sunlight and are subject to periodic desiccation.

Procedure:

1. The life cycle of a typical fern is shown in **Figures 3-6 and 3-7.**

2. Examine living fern gametophytes for antheridia and archegonia. Motile sperm are likely to be seen.

3. Prepared slides are also available. Also look at the sori from the underside of a fern leaf at 10X magnification.

Plants with Seeds

Background (Gymnosperms)

The geological record indicates that during certain periods of the earth's history, gymnosperms (vascular plants whose seeds are not enclosed by ovarian tissue) were relatively abundant. Some classes of gymnosperms are now extinct and known only from their fossil remains. One class is represented by a single surviving species, the ginkgo tree. (Near Alderman Library or in the Engineering School quadrangle.) The largest classes are the Cycadae, which are relatively rare plants of tropical and subtropical regions, and the Coniferae, which include the pines, spruces, and related shrubs that are mainly inhabitants of the temperate regions.

While doubtlessly reduced in number from former times, more than 600 species are extant today. Most of these are trees; some, like the redwood and sequoia, are giants towering up to 300 feet. Others, such as many of the junipers, are shrubs. In general, the species are evergreen, but several, like the bald cypress and larches, lose their leaves in the winter.

Because the group is so large and highly diversified, the morphology is exceedingly varied.

Coniferophyta (Conifers)

From an economic standpoint, conifers are of great importance. They are used widely as a source of lumber in building and inexpensive grades of furniture. Spruce and southern pine are used extensively in the manufacture of wood pulp. Spruces and firs together constitute the great bulk of the evergreens sold annually as Christmas trees. Telephone poles, railroad ties, and lumber for crates and boxes are for the most part made of conifers.

The pine will serve as a typical representative of this group in the laboratory. Pine trees with their roots, stems, and leaves (needles) are the mature sporophytes. Review the life cycle of the pine illustrated in **Figure 3-8**.

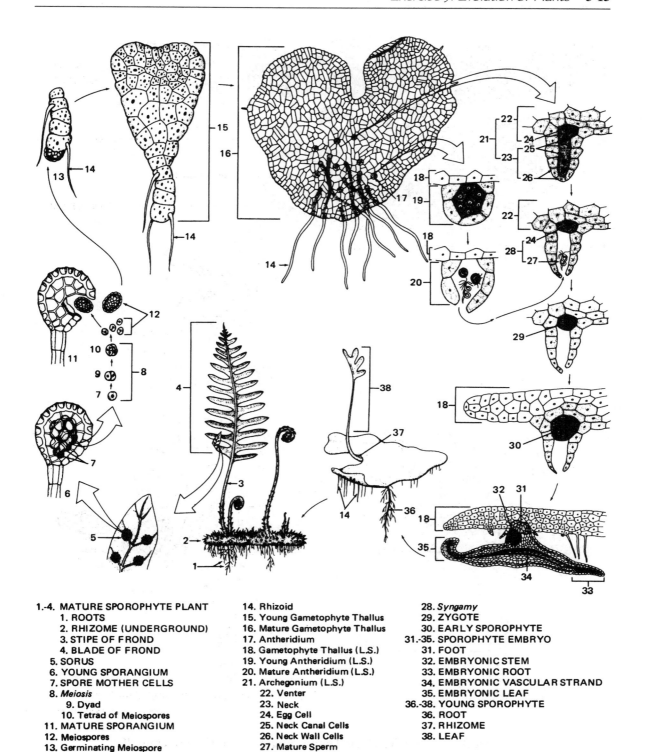

1.-4. MATURE SPOROPHYTE PLANT
    1. ROOTS
    2. RHIZOME (UNDERGROUND)
    3. STIPE OF FROND
    4. BLADE OF FROND
5. SORUS
6. YOUNG SPORANGIUM
7. SPORE MOTHER CELLS
8. *Meiosis*
    9. Dyad
    10. Tetrad of Meiospores
11. MATURE SPORANGIUM
12. Meiospores
13. Germinating Meiospore

14. Rhizoid
15. Young Gametophyte Thallus
16. Mature Gametophyte Thallus
17. Antheridium
18. Gametophyte Thallus (L.S.)
19. Young Antheridium (L.S.)
20. Mature Antheridium (L.S.)
21. Archegonium (L.S.)
    22. Venter
    23. Neck
    24. Egg Cell
    25. Neck Canal Cells
    26. Neck Wall Cells
    27. Mature Sperm

28. *Syngamy*
29. ZYGOTE
30. EARLY SPOROPHYTE
31.-35. SPOROPHYTE EMBRYO
    31. FOOT
    32. EMBRYONIC STEM
    33. EMBRYONIC ROOT
    34. EMBRYONIC VASCULAR STRAND
    35. EMBRYONIC LEAF
36.-38. YOUNG SPOROPHYTE
    36. ROOT
    37. RHIZOME
    38. LEAF

Figure 3-6. Life Cycle of a Fern

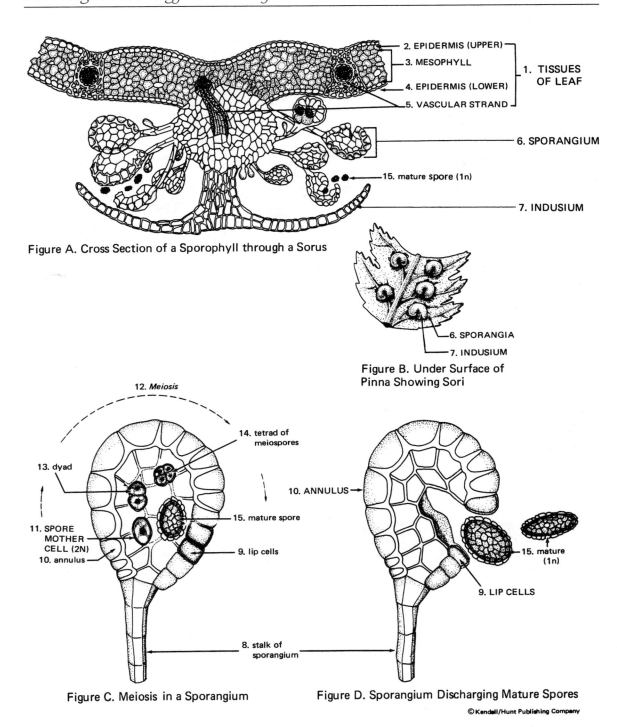

Figure A. Cross Section of a Sporophyll through a Sorus

2. EPIDERMIS (UPPER)
3. MESOPHYLL
4. EPIDERMIS (LOWER)
5. VASCULAR STRAND
1. TISSUES OF LEAF

6. SPORANGIUM

15. mature spore (1n)

7. INDUSIUM

6. SPORANGIA
7. INDUSIUM

Figure B. Under Surface of Pinna Showing Sori

12. *Meiosis*

14. tetrad of meiospores

13. dyad

10. ANNULUS →

15. mature spore

11. SPORE MOTHER CELL (2N)
10. annulus

9. lip cells

15. mature (1n)

9. LIP CELLS

8. stalk of sporangium

Figure C. Meiosis in a Sporangium

Figure D. Sporangium Discharging Mature Spores

Figure 3-7. Fern Life Cycle Detail. Structure of the Sporophyll

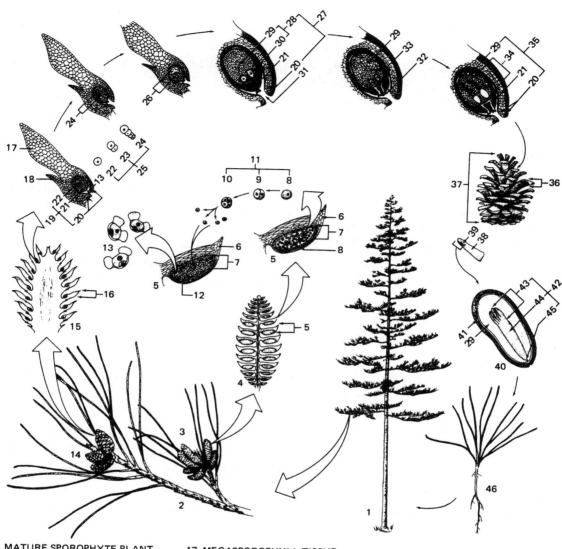

1. MATURE SPOROPHYTE PLANT
2. BRANCH WITH CONES
3. STAMINATE CONES
4. STAMINATE CONE (L.S.)
5. MICROSPOROPHYLL
6. MICROSPOROPHYLL TISSUE
7. MICROSPORANGIUM
   8. MICROSPORE MOTHER CELL
   9. Dyad
   10. Tetrad of microspores
   11. *Meiosis*
12. Young pollen grain
13. Pollen grains
14. OVULATE CONE
15. OVULATE CONE (L.S.)
16. MEGASPOROPHYLL

17. MEGASPOROPHYLL TISSUE
18. BRACT
19. OVULE
   20. INTEGUMENTS
   21. NUCELLUS (MEGASPORANGIUM)
   22. MEGASPORE MOTHER CELL
   23. Dyad (L.C.)
   24. Tetrad of megaspores
   25. *Meiosis*
26. YOUNG FEMALE GAMETOPHYTE
27. MATURE OVULE
   28. Mature female gametophyte
   29. Female gametophyte tissue
   30. Archegonium with egg cell
   31. Micropyle

32. Germinated pollen grain
33. ZYGOTE (After *syngamy)*
34. YOUNG EMBRYO
35. MATURING SEED
36. MEGASPOROPHYLL
37. MATURE OVULATE CONE
38. WING OF SEED
39. SEED
40. SEED (L.S.)
41. SEED COAT
42. EMBRYO
   43. COTYLEDONS
   44. HYPOCOTYL
   45. RADICLE
46. YOUNG SPOROPHYTE

Figure 3-8. Life Cycle of the Pine

The pines, spruces, yews, firs, and so on are characterized by woodiness and by scale or needlelike leaves. These plants introduce the method by which trees increase in girth, which is an index of their cambial activity. Conifers produce two kinds of spores. Microspores develop into microgametophytes (pollen grains), which produce male gametes, and megaspores develop into megagametophytes, which produce several archegonia, each of which contains a female gamete. Both kinds of gametophytes are reduced in size and structural complexity and are dependent on the sporophytes. The sporophytes of most species are large evergreen trees, shrubbery, and arborescent perennials. All conifers produce seeds that house the plant embryo.

Some conifers are dioecious, but more are monoecious. Conifers are have several adaptations for life on land. The female gametophyte is retained until after fertilization. The female gametophyte and surrounding sporophyte tissues are properly called the ovule before fertilization, and the seed after fertilization. In addition to this protection immediately around the female gametophyte, a woody structure protects the aggregation of ovules (or seeds). This is the familiar pine cone. The sporophytes bear two kinds of woody cones—the small pollen cones (males) and larger seed cones (females).

Procedure

1. Examine a preserved male cone. Note overlapping microsporophylls. Remove one and examine the two microsporangia. Crush one of the microsporangia and make a wet mount for microscopic study. Each pollen grain is an immature microgametophyte. Examine prepared longitudinal slides of a **male cone** that show the pollen grains. Within the pollen tube is a large tube nucleus that will direct the synthesis of the pollen tube. The generative cell nucleus will also be present and will form 2 sperm.

2. Examine preserved female cones. Remove a complete megasporophyll (scale) to observe the two ovules located on the upper side, near the center of the cone. Understand that what you are viewing is shown on **Figure 3-8** (Life Cycle of Pine).

3. Prepared slides: ovulate cone and male cone (refer to **Figure 3-9** for help).

4. You are not responsible for knowing the entire pine life cycle. Understand how the parts of the pine cone contribute to the overall life cycle of the pine.

Background (ANTHOPHYTA or ANGIOSPERMS)

The flowering plants are the most common plants on the earth's land surface. They are recognized by their broad leaves and their reproductive feature of having flowers. Angiosperms differ from the gymnosperms in that their ovules and seeds are enclosed within a carpel or megasporophyll that later becomes a seedbearing fruit. This not only helps protect the seeds but also aids in dispersal.

The process of double fertilization occurs only in the angiosperms. The gametophytes of angiosperms are reduced even more than those of the gymnosperms. The megagametophytes of

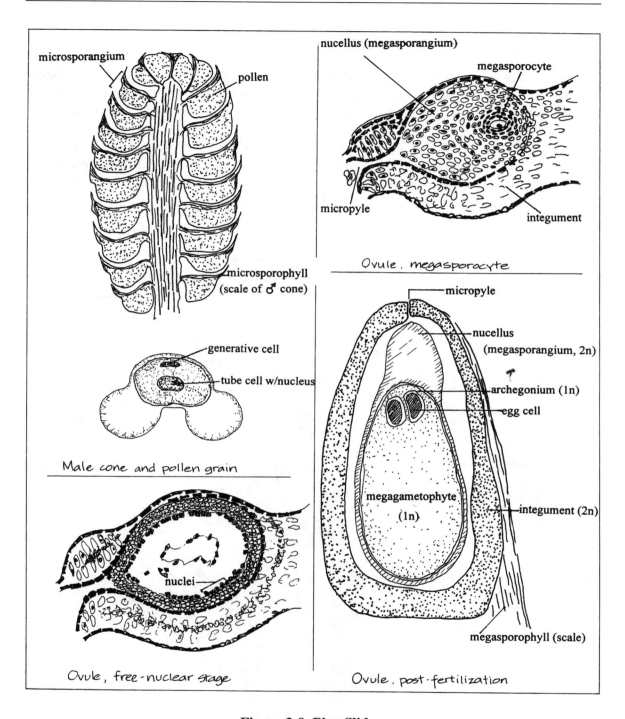

Figure 3-9. Pine Slides

angiosperms never produce archegonia. It has been estimated that there are approximately 250,000 species of plants that reproduce by means of flowers and no two species have exactly the same floral structure. Morphologically, a flower resembles a leafy shoot whose stem has been greatly modified, and bears highly specialized leaves and branches referred to as floral organs.

The Flower

Basically, there are four kinds of floral organs: <u>sepals</u>, <u>petals</u>, <u>stamens</u>, and <u>carpels</u> (**Figure 3-10**). Flowers have one or more stamens or carpels, which are the floral organs involved directly in reproduction. The petals and sepals may be present or absent, as they are not essential.

Today we will examine a flower and note some of the important floral structures. Fundamental knowledge of floral structure is a prerequisite to understanding the process of reproduction in flowering plants, which will also be considered in today's exercise. Floral organs are attached to an axis in whorls or spirals. The outermost whorl is composed of the leaf-like

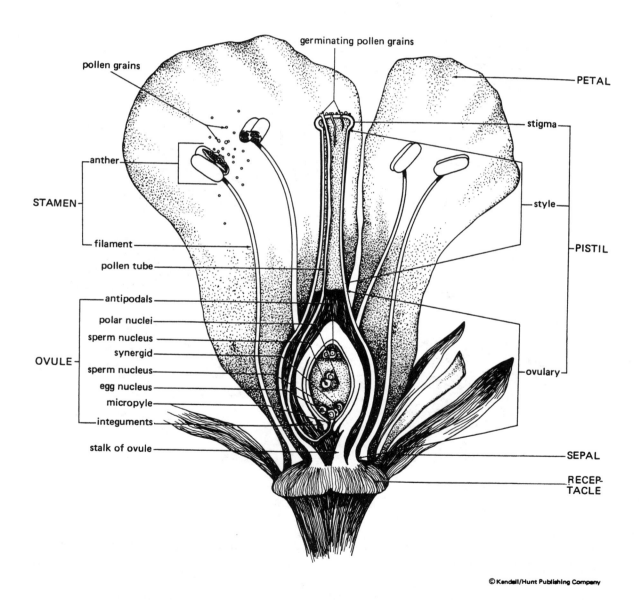

Figure 3-10. Generalized Structure of a Dicot Flower

sepals. The second is ordinarily composed of petals. The sepals are collectively called the calyx, and the petals are collectively called the corolla. The petals are usually thin in texture and colored. The third group of parts is composed of the stamens, often arranged on an elongated receptacle in a spiral fashion. The innermost whorl consists of the pistil. The stamen is composed of a stalk-like filament that is attached to the pollen-producing anther at its upper end. A cross section of an anther is shown in **Figure 3-11**. Each anther has four lobes arranged in two pairs. Each lobe contains a cavity or locule. Within the locule of a young anther, there are a large number of microspore mother cells or microspores (if meiosis has occurred). At the end of the first meiotic telophase, a wall is formed between the two cells. These two cells are called the dyad. At the end of the second meiotic telophase, four haploid cells are formed and these cells cling together for awhile to form a tetrad. The members of tetrads eventually separate and grow somewhat in size to form the uninucleate microspores.

The single nucleus of each microspore divides again. Usually these two nuclei are not separated by a cross wall. A thick, very often sculptured layer of cutin is deposited on the cell wall and the resulting structure is a pollen grain. The two nuclei in the pollen grain will have a different function in the further development of the microgametophyte. One of them is the tube nucleus and the other is the generative nucleus. The generative nucleus is sometimes distinguishable by an observable differentiation in the cytoplasm.

After the pollen is shed to a receptive stigma, pores in the wall of the pollen grain permit the exit of a pollen tube that grows into the style. The tube nucleus migrates into the pollen tube through the style and into the ovary, where it enters an ovule by way of the micropyle. While the pollen tube is growing through the style, the generative nucleus divides to form two sperm cells. This structure consisting of a pollen grain, pollen tube cell with tube nucleus, and two sperm cells is the mature microgametophyte. The central pistil has three divisions: A swollen ovary at the base, a style and the stigma. The ovary contains one or more ovules, which mature into seeds after fertilization.

Note that the mature gametophyte is reduced to seven nuclei in the embryo sac. The organization of the embryo sac and the angiosperm life cycle is shown in **Figure 3-12**. Test your knowledge of the flower cycle with the key for terms.

Zygote Formation

The development of both the male (micro) and female (mega) gametophytes has been traced to maturity. The next critical stage in the life cycle is the union of the male (sperm) and female (egg) gametes to form a zygote, which is the first cell of the new sporophyte. After a pollen tube has grown through the style and into an ovule by way of the micropyle, it penetrates the embryo sac and ruptures. The tube nucleus and two sperm nuclei are discharged into the embryo sac. Both sperm, you will recall, are haploid, or 1n. One sperm fuses with the egg, also haploid, to form a zygote, which is diploid, or 2n. The other sperm fuses with the polar nuclei to form a single nucleus, the triploid endosperm nucleus. The process whereby two sperm unite with two different nuclei in the megagametophyte is known as double fertilization. It is unique in the angiosperms. The tube nucleus, the synergids, and the antipodals all eventually degenerate or disintegrate and normally play no part in the development of the zygote or endosperm.

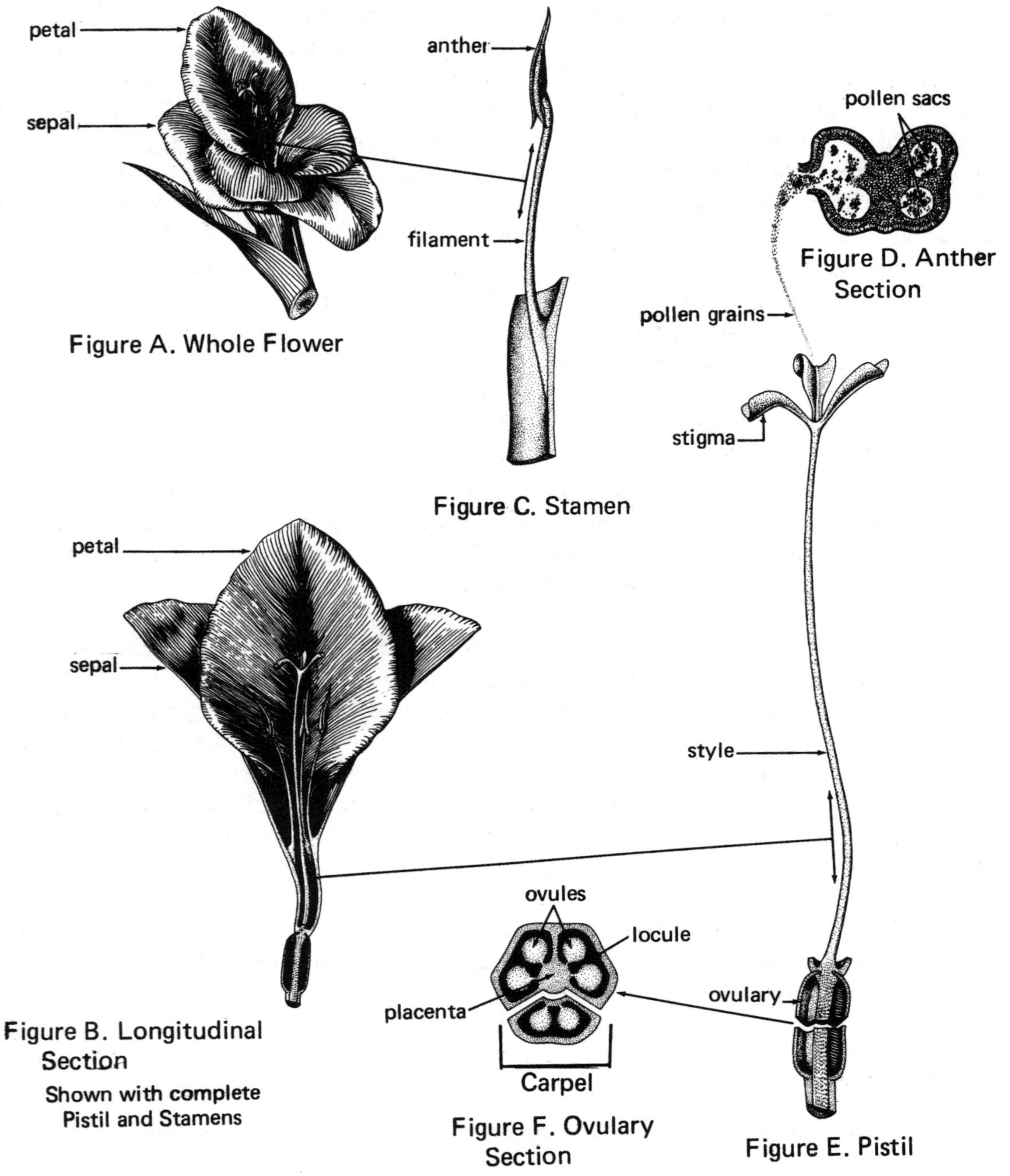

petal

sepal

Figure A. Whole Flower

anther

filament

Figure C. Stamen

pollen sacs

Figure D. Anther Section

pollen grains

stigma

petal

sepal

style

Figure B. Longitudinal Section
Shown with complete Pistil and Stamens

ovules

locule

placenta

ovulary

Carpel

Figure F. Ovulary Section

Figure E. Pistil

Figure 3-11. Structure of a Monocot Flower

The young sporophyte or embryo thus draws nourishment from the endosperm, which in turn drew its nourishment from the parent sporophyte. There is no good reason why the young embryo could not draw its nourishment directly from the 2n tissue around it, without the intermediary endosperm. Experiments have shown, however, that the relationship of the endosperm to the nucellus is crucial to the development of the embryo. One might think that double fertilization, yielding an embryo and endosperm, is too obscure and economically unimportant to be worth emphasis. In fact, endosperm is the world's most important single source of food. It is the starchy part of all grains that constitutes the bulk of the calories in civilization's three greatest foodstuffs: rice, wheat, and corn.

The zygote now develops into an embryo that grows at the expense of the endosperm. Note that the ovary and ovule increase greatly in size. The embryo is located at the micropylar end. Its cells are smaller and stain a little darker than those in the endosperm and ovary wall. The endosperm surrounds the embryo and completely fills the remainder of the embryo sac. Some nucellar tissue may still be present, but it has started to disintegrate. The integument can be seen but it is relatively thin in proportion to the other parts of the ovule.

Plant embryos develop through a sequence of well-defined stages (**Figure 3-13**). These embryos will likely be at the socalled heart or torpedo stage. They will not live outside of the plant, although they can be grown in sterile culture on a nutrient medium that contains inorganic salts and a carbohydrate source.

The flower is an evolutionary expression that extends back unknown millions of years and continues today. It is unique and its success is shown by the dominance of flowering plants throughout the world. As flowers have evolved, extraordinary relationships have occurred to assure reproductive successes. We end this exercise with a video documenting this wonderful story.

Procedure

1. Under the dissecting microscope, note the individual parts of the flower. Begin from the outside of the flower and locate the calyx, then the corolla. Remove these portions so that the essential organs can easily be identified such as the stamen and pistil.

2. Prepare a wet mount of a cross section of an ovary. Compare it to a prepared slide and Figure 3-11. Locate the locules and the developing ovules.

3. With instructions from the TA, **dissect embryos from a lima bean.**

Background (Carnivorous plants)

Dionea muscipula is the famous Venus Flytrap, and it is the only species in its genus. This rare plant is found in southern North Carolina and the bordering South Carolina. Insect capture is performed by attracting the insects with nectar to bilobed leaves, which snap shut upon the prey.

Key to Figure 3-12

1. MATURE SPOROPHYTE
2. FLOWER
   3. SEPAL
   4. PETAL
   5. STAMEN
      6. ANTHER
      7. FILAMENT
   8. PISTIL
      9. STIGMA
      10. STYLE
      11. OVULE
      12. OVULARY
13. YOUNG ANTHER (C.S.)
14. MICROSPORANGIUM
15. *Meiosis*
    16. MICROSPORE MOTHER CELL
    17. Dyad
    18. Tetrad of Microspores
19. MATURING ANTHER (C.S.)
20. MATURE ANTHER (C.S.)
21. Pollen grains (male gametophyte)
22. YOUNG OVULE (L.S.)
    23. MICROPYLE
    24. INTEGUMENTS
    25. NUCELLUS (MEGASPORANGIUM)
    26. MEGASPORE MOTHER CELL
27. *Meiosis*
    28. Dyad
    29. Linear tetrad of megaspores
30. Functional megaspore
31. Disintegrating megaspores
32. *Mitosis* (Functional megaspore)
33. Young embryo sac (female gametophyte)
34. Haploid nuclei
35. 4-Nucleate embryo sac

36. 8-Nucleate embryo sac
37. Mature ovule
38. Mature embryo sac
    39. Synergids
    40. Egg cell
    41. Polar nuclei
    42. Antipodal nuclei
43. *Double fertilization*
    44. Germinated pollen grain
    45. Pollen tube
    46. Egg cell
    47. Sperm nucleus
    48. Polar nuclei
    49. Sperm nucleus
    50. Antipodals
51. INTEGUMENTS OF OVULE
52. OVULARY WALL
53. ZYGOTE *(After Syngany)*
54. ENDOSPERM NUCLEUS (3N)
55. Deteriorating antipodals
56. YOUNG SPOROPHYTE EMBRYO
57. ENDOSPERM TISSUE (3N)
58. YOUNG FRUIT
    59. EMBRYO DIFFERENTIATED
    60. YOUNG SEED
    61. WALL OF FRUIT
62. REMNANTS OF STYLE
63. MATURE FRUIT
64. SEED
65. SEED COAT
66. EMBRYO
67. PERICARP (FRUIT WALL)
68. SEED GERMINATING
69. YOUNG SPOROPHYTE

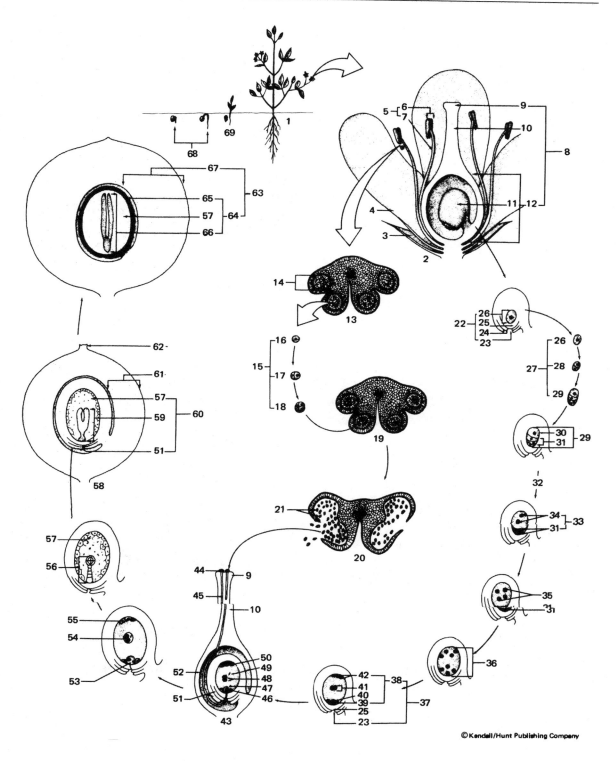

Figure 3-12. Life Cycle of an Angiosperm

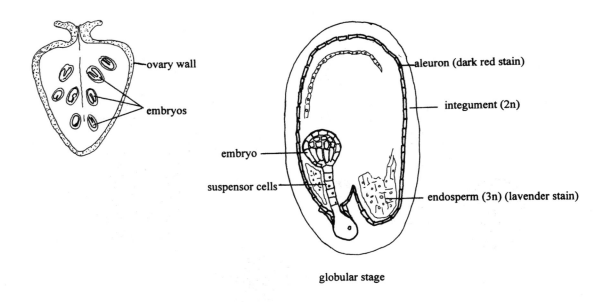

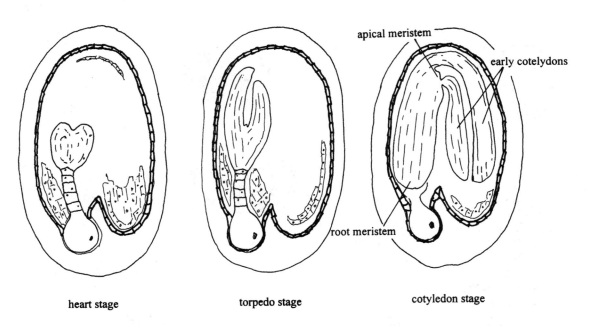

Figure 3-13. Capsella Ovary, Cross-Section, Showing Stages of Embryo Development

Drosera is also called the sundew. There are about 150–160 Drosera species described, and they are scattered around the globe. These plants bear stalks or tentacles on their leaves, and these stalks are tipped with glands (which are often brightly coloured). The glands exude attractive nectar, adhesive compounds, and digestive enzymes. Insects that land on the leaves stick fast and are digested. Often nearby glandular tentacles are stimulated and also adhere to the insect, and on many species the entire leaf coils around the prey. These motions are slow, taking minutes or hours to occur, although some species (for example *D. burmannii*) exhibit faster motion.

Nepenthes are called pitcher plants. These are the long vines that snake through the undergrowth and trees in the jungles of Southeast Asia, northern Australia, and Madagascar. The pitchers, borne at the tips of the leaves on tendrils, are shaped like tubes, tubs, or drums, and capture various small invertebrates (and the occasional small vertebrate).

Similar to *Nepenethes* are *Sarracenia*. *Sarracenia* have no rapidly moving parts. They capture insects by not letting them escape from their pitchers once the prey topple in. Some species exude an amazing variety of chemicals, including digestive enzymes, wetting agents, and insect narcotics! Others rely on bacterial action to digest the prey.

Procedure

1. Observe the different strategies used by the different species of carnivorous plant.

2. Your TA will demonstrate the Venus Flytrap for the class.

TA Notice

The last 30 minutes should be reserved for the video. This exercise concludes with a portion of the film Sexual Encounters of the Floral Kind.

Review Questions

1. The earliest fossil record of vascular plants is how long ago?

2. The first known land plant was <u>Cooksonia</u>. Describe it.

3. Regarding <u>Psilotum</u>:

    (a) What division?

    (b) What are the trilobed structures on the stem?

4. What feature unites <u>Equisetum</u> and <u>Calamites</u>?

5. Developing pollen grains on male cones are contained within _____ .

6. The gametophyte generation is found inside the sporophyte for the first time in _____ .

7. Gametophyte or sporophyte?

   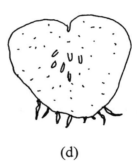

    (a)   (b)   (c)   (d)

8. The end products of double fertilization are _____ and _____ .

9. The ovary develops into _____ ; the ovule into the _____ .

Answers

1. 438–408 mya

2. Less than 50 cm. tall without true leaves or roots. Cooksonia and other early plants looked much like Psilotum.

3. a. Psilophyta   b. sporangia

4. ribbed stem, whorls of leaves

5. microsporangia

6. gymnosperms

7. a. gametophyte
    b. sporophyte
    c. sporophyte
    d. gametophyte

8. zygote and endosperm

9. fruit, seed

Reference

Foster, A.S. and E.M. Gifford, Jr. 1974. Comparative Morphology of Vascular Plants. W.H. Freeman.

Bio 204 Review Checklist

Phylum CHLOROPHYTA	*Volvox*	colonial	flagellate
	Spirogyra	multicellular pyrenoid	conjugation tube spiral chloroplast
	Scenedesmus	colonial (4–8)	pyrenoid
	Micrasterias	desmid	isthmus
Division Bryophyta	*Hepaticae*	liverwort gemmae archegonial receptacle anteridial recetacle embryophyte condition Egg Spores	gametophyte gemmae cup sperm diecious
	Musci	moss thallus Calyptra Rhizoids Capsule spores	diecious operculum egg, sperm gametophyte*
Division PSILOPHYTA	*Psilotum*	no leaves/roots homosporous*	epiphyte trilobed sporangium
Division LYCOPHYTA	*Selaginella*	heterosporous* megaspore microspore true leaves* *Lycopodium* *Lepidodendron* (fossil)	strobilus megasporangium microsporangium sporophylls homosporous*
Division SPHENOPHYTA	*Equisetum*	nodes/internodes ribbed stems homosporous*	whorled leaves microphylls strobilus

Division PTEROPHYTA	(several ferns)	antheridia	archegonia
		embryo	rhizoids
		sorus	indusium
		annulus	
		homosporous*	spores
		rhizome	lip cells
		dominant sporophyte*	

| Division CONIFEROPHYTA | *Pinus* | staminate cone | pollen grains |
| | (Gymnosperms) | generative nucleus | tube nucleus |

Division ANTHOPHYTA	*Geranium,*	calyx	corolla
(Angiosperms)	lima bean	stamen	filament
		anther	ovary
		ovule	pistil
		style	stigma
		carpel	locule
		sepals	endosperm
		double fertilization*	polar nuclei
		heart-shaped embryo	globular embryo
		cotyledons	suspensor cells
		fruit*	seed

	Dionaea	Note adaptions!	
	Drosera		
	Nepenthes		
	Sarracenia		

The Animal Kingdom Lower Invertebrates

Introduction

The multicellular animal is recorded in fossils over 600 million years old. The Precambrian seas contained many species, all of them with one important characteristic—they were ingesters. Incapable of photosynthesis like algae and plants, nourishment came from ingesting other organisms. The quest, capture and management of food is a central driving force in the evolution of animals and the diversity of our animals species demonstrates how common problems for survival have been answered in different, but successful ways.

Over 90 percent of our animal species are invertebrates—that is, animals without a backbone or vertebrate column. Our purpose in the lab will be to survey representative animals of a few invertebrate phyla, thereby permitting more time for detailed study.

Phylum Porifera

Background

Most sponges are marine, but one family does occur in fresh water. The adults are attached but have free-swimming larval stages. An individual animal typically consists of a central cavity surrounded by a body wall pierced by many small canals that connect the central cavity to the outside (**Figure 4-1**). An opening, the <u>osculum</u>, at the non-attached end of the animal, permits the passage of water and other materials from the central cavity. The body wall contains a <u>skeletal system</u> of <u>spicules</u> of various shapes. Flagellated collar cells (<u>choanocytes</u>) are present in all the sponges.

Procedure

1. Examine the sponges on demonstration.

2. Obtain a piece of <u>Grantia</u> and use your dissecting microscope to search for tiny pores (ostia) through which water is taken into the body cavity (<u>spongocoel</u>). Make a thin section water mount and examine under 10X. Apopyles can be seen lining the inside wall of the body cavity.

3. In a prepared slide of Grantia, identify the choanocytes lining the spongocoel. What is their function? What separates the two body layers?

4. A drawing is not required. Study **Figure 4-1** and relate it to what you have observed.

Phylum Cnidaria (Coelenterata)

Background

Most cnidarians are marine, but some occur in fresh water. The members of the phylum are radially symmetrical. The cnidarian body is formed from an outer layer of ectoderm and an inner layer of endoderm, separated by a noncellular layer of jellylike mesoglea. All members of the phylum have a gastrovascular cavity (coelenteron), which is a digestive cavity with only one opening. Surrounding the coelenteron are the two layers of cells derived from two embryonic germ layers; thus, the animals are referred to as diploblastic. The unique characteristic of the phylum Cnidaria is that all members possess tentacles with nematocysts. These are characteristic stinging organelles with a complex structure. Some common examples of colonial cnidarians are coral, Obelia, and the Portuguese man-of-war. In these colonies, individual animals have different forms and structures adapted for specialized functions. Such colonies with individuals of more than one body form exhibit polymorphism. The attached vegetative cnidarian types are called polyps, and the free-swimming, sexually reproducing forms are called medusae. Cnidarians are divided into the classes discussed in the following sections. We will observe members of the class Hydrozoa, Scyphozoa, and Anthozoa.

Class Hydrozoa

Hydrozoa are chiefly polyps, but some have medusae that are small, and a velum. This cnidarian is not divided by a septum. Obelia and Physalia are marine colonial forms. Hydra are essentially sessile organisms; however, they can change locations by floating. Another method of moving about is by a means of locomotion similar to somersaulting. The Hydra can bend its tentacles and oral end over, grasp onto the substrate, and then release its foot. The Hydra then bends its foot over and adheres to the next substrate, thus changing locations.

Hydra feeds primarily on small, living crustaceans, waiting for its food to come to it. Food is captured with the tentacles and taken into the gastrovascular cavity through the mouth. The capture of prey is also aided by the presence of nematocysts (stinging cell) on the surface of the tentacles. The nematocysts are produced by and held within a cell called a cnidoblast. At 40x, you may be able to see the cnidocil, which is the trigger. Under ordinary circumstances the discharge of the nematocysts is triggered by physical contact with the prey, but any sort of rough treatment will accomplish the same effect. The body wall of the Hydra is made of two cell layers, the epidermis and gastrodermis. The epidermis is composed of five cell types. Epithelio-muscle cells form most of the surface of the epidermis. Smaller interstitial cells are found beneath the epidermal surface between the epithelio-muscle cells. They are the germinal cells of the Hydra. Cnidoblasts are located throughout the epidermis. Mucous-secreting cells are abundant near the basal foot area. Sensory cells are scattered throughout the epidermis. Nerve cells are located at the base of the epidermis next to the mesoglea. The gastrodermis is composed of several types

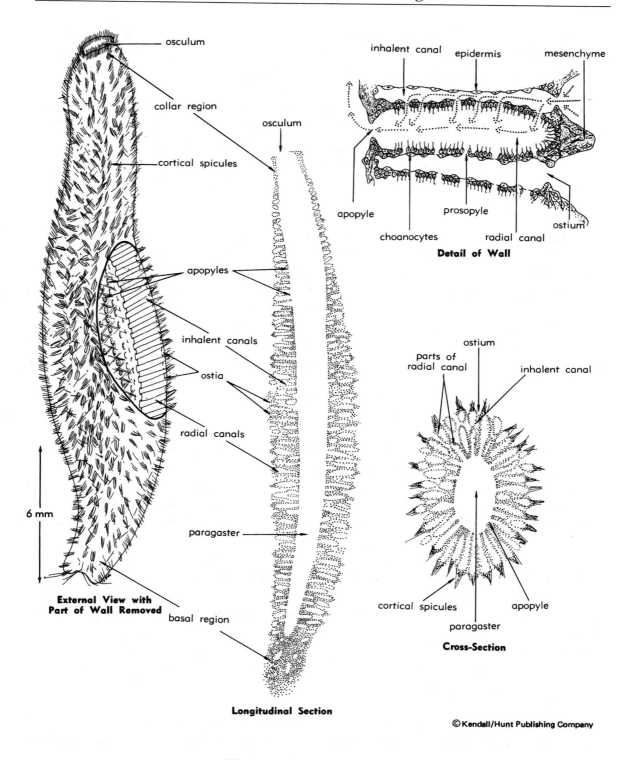

Figure 4-1. Sponge Anatomy

of cells. Enzymatic-gland cells and epithelial-digestive cells have already been discussed. Nerve cells are present but in a smaller number than in the epidermis. Mucous-secreting cells are prevalent near the mouth. Gas exchange occurs by diffusion through the body surface.

Once the prey is taken in through the mouth of the <u>Hydra</u> it enters its gastrovascular cavity. <u>Enzymatic-gland cells</u> in the lining of the gastrovascular cavity secrete digestive enzymes, and flagellated cells aid in mixing. <u>Epithelial-digestive cells</u> engulf small pieces of food, and digestion continues in food vacuoles within these cells. Digested food products diffuse throughout the body cells. Indigestible materials are forced out through the mouth. Nitrogenous wastes diffuse through the body surface.

Within the class Hydrozoa most members have a life cycle in which the vegetative polyp produces medusae that are then freed. The medusae grow and gonads develop within them. These gonads produce eggs and sperm. From the fertilized egg, a polyp develops and the cycle starts over again. <u>Obelia</u> is an example of a hydrozoan with this type of life cycle. In other members of the class Hydrozoa, the medusae are produced on the polyp but are not freed. The gonads then develop in the medusae as they are attached to the polyp. In <u>Hydra</u> and some others, medusae do not develop. The gonads develop on the polyp itself, the testis on the upper half of the body and the ovary on the lower half. <u>Hydra</u> is capable of sexual reproduction by budding, but does not rely exclusively on this for its propagation. If the bud remains attached to the parent, the result will be a colonial hydroid, a form commonly encountered in marine cnidarians. In <u>Hydra</u> the buds drop off to live a solitary life.

<u>Obelia</u> is a typical colonial marine cnidarian common in shallow water and on piles, rocks, and seaweed. The "branches" of the colony terminate in two types of polyps: feeding polyps and reproductive polyps. The <u>feeding polyps</u> (hydranths) can be identified by the presence of tentacles bearing cnidoblast cells capable of releasing nematocysts for the capture of prey. The food gathered by these polyps enters their <u>gastrovascular cavity</u>. Notice that this hollow internal cavity continues through the branches that support all the polyps. In this way the gastrovascular cavity serves as both the site of digestion and as the means of distributing to the entire colony the food procured by the feeding polyps. Next identify a <u>reproductive polyp</u> (gonangium); it is club-shaped and lacks tentacles. Within this polyp is a blunt projection to which are attached a number of medusoid buds. These buds will eventually be released from the opening at the end of the polyp and swim away as tiny medusae. When they are mature, the medusae will produce eggs or sperm and release the gametes into the sea where fertilization takes place. The zygote will develop into a free-swimming ciliated planula larva, which will eventually settle on a solid surface and develop into a new colonial hydroid generation.

We will also take a look at a Portuguese man-of-war (<u>Physalia</u>). Some of the polyps possess tentacles with nematocysts. Others are digesters for the colony. Still others are specialized to produce eggs and sperm. The most obvious individual in the colony is the modified medusa that forms the float. <u>Physalia</u> is well known to swimmers. The venomous nematocysts can kill small fish and their sting is very painful to humans.

Class Scyphozoa

The class Scyphozoa comprises the jellyfish. In this class, the medusa is the dominant stage and the polyp is relatively inconspicuous. All members of the class are marine and bell-shaped, and generally do not have a velum, an inward projection forming a shelf at the margin of the bell.

Aurelia is one of the most common jellyfish; it is found throughout the world. These jellyfish are often seen in large shoals drifting or swimming slowly by rhythmically contracting the saucer-shaped bell. The medusa is the form of Aurelia most commonly observed, although it undergoes a life cycle similar to Obelia in that the medusa and polyp forms alternate.

Class Anthozoa

The class Anthozoa comprises the sea anemones and corals. Members of this class are all polyps, and there is no medusa stage. There are many colonial forms, all marine. Metridium (**Figures 4-7 and 4-8**) and Corallium are common forms of this class.

Procedure (Class Hydrazoa)

1. Obtain a living Hydra for observation. Record your study by sketches and notes (see **Figures 4-2–4-4**).

    a. **Form.** Hydra is a freshwater polyp without a medusa stage. With a hand lens or the dissecting microscope, observe your specimen. Note the characteristic body form. Note the sac-like tube with a conical projection at one end and the hypostome surrounding the mouth. Around the base of the hypostome are the flexible tentacles. Count the number of tentacles. What function do the tentacles have? Is the basal end of Hydra attached or does it wave freely? The foot or the basal disc is one of the specialized cell groups found in this animal. The cells making up this tissue are capable of secreting a mucous, adhesive material.

    b. **Feeding.** Before Hydra will accept food, **it must be left undisturbed for a considerable time,** and must be in a fully expanded condition. Therefore place your sample of Hydra in a small dish on the stage of the dissecting scope **with the light off.** Be careful not to disturb or bump the dish. Brine shrimp will be added to your dish to allow you to observe feeding in Hydra.

    c. **Life Cycles and Sexual Reproduction.** Prepared slides will be available showing the sexual forms of Hydra. You may want to compare them to **Figure 4-2**.

    d. **Asexual Reproduction and Regeneration.** Observe the prepared slide of the budding Hydra, an individual that has sprouted progeny from the side of its tubular body.

    e. **Body Wall.** Prepared slides of transverse sections through the tubular body of Hydra should be viewed first with the low, and then with the high-power lens on the compound microscope. Note the outer and inner layers of cells with a thin mesoglea between (**Figure 4-3**). The cavity in the center is the gastrovascular cavity, which extends up into the tentacles. In most of the slides available to you, the nuclei are stained blue and purple and appear granular. Oval objects lying in the cytoplasm on the outer layer of cells,

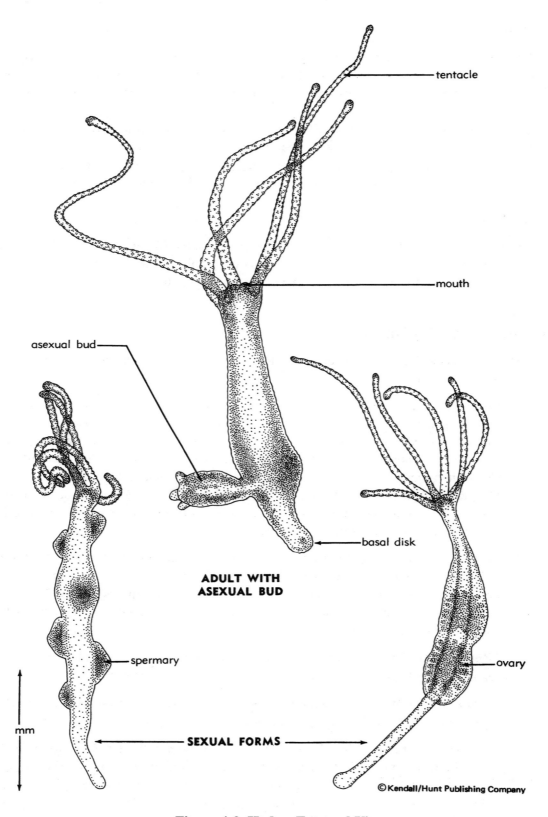

tentacle

mouth

asexual bud

basal disk

**ADULT WITH
ASEXUAL BUD**

spermary

mm

ovary

SEXUAL FORMS

© Kendall/Hunt Publishing Company

Figure 4-2. Hydra, External View

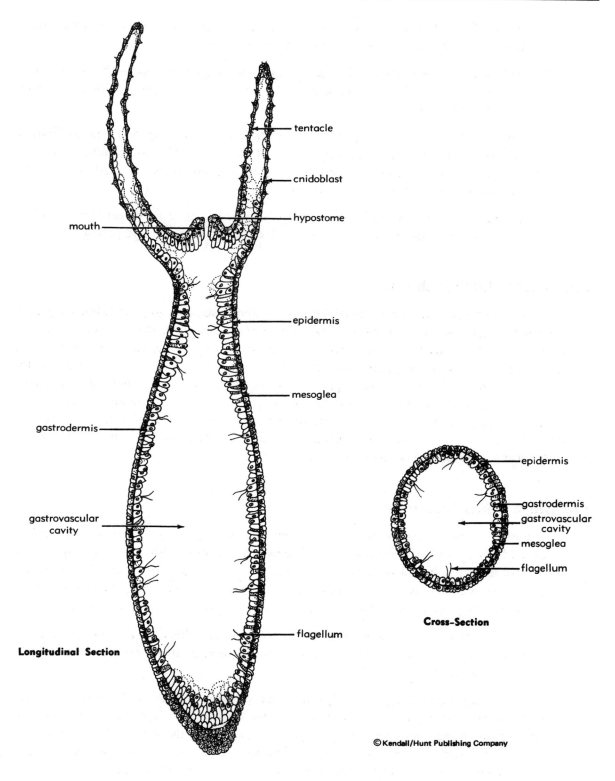

tentacle

cnidoblast

hypostome

mouth

epidermis

mesoglea

gastrodermis

gastrovascular
cavity

flagellum

Longitudinal Section

epidermis

gastrodermis

gastrovascular
cavity

mesoglea

flagellum

Cross-Section

Figure 4-3. Hydra, Sections

and equal to or larger than the nuclei, are probably nematocysts. These are particularly numerous in the tentacles, but are also present on the body of <u>Hydra</u>.

 f. **Discharge of nematocysts.** Placing the a <u>Hydra</u> on a microscope slide in a drop of dilute safranin solution and add a coverslip. Observe such a preparation under high power (**Figure 4-4**).

2. Examine a prepared slide of an <u>Obelia</u> colony under the microscope and refer to **Figures 4-5 and 4-6.** Notice the plant-like appearance of the colony. The "stem" and the polyps have a transparent covering, the <u>perisarc</u>, which functions to protect and support the colony, and is easily distinguished on the slide.

3. Examine the preserved specimen of a Portuguese Man-o-war on demonstration. This is a free-living, complex hydrozoan colony composed of both medusae and polyps.

Procedure (Class Scyphozoa)

Examine the preserved specimens of <u>Aurelia</u> on demonstration in your laboratory and compare with Figure 4-7. Notice that the body is umbrella-shaped and that the rim of the bell is fringed by a row of closely spaced tentacles. Sense organs are very useful to an organism as mobile as <u>Aurelia</u>. It is not surprising then that the tentacles are interrupted by eight equally spaced indentations, each with a sense organ. The sense organs consist of a pigmented eyespot, sensitive to light; a hollow <u>statocyst</u>, containing minute particles whose movements set up stimuli that detect swimming movements and positions; and two sense pits, lined with cells that are thought to be sensitive to food or other chemicals in the water. The mouth is found in the center of the concave surface and opens into the gastrovascular cavity.

Procedure (Class Anthozoa)

1. Examine the preserved specimens of the sea anemone <u>Metridium</u>. Note the stout cylindrical body expanded at the upper end. The mouth is in the center surrounded by several rows of tentacles. It leads to a gullet and then to the gastrovascular cavity. The basal end forms a smooth muscular disc on which the anemone can slowly move about or by which it can hold onto rocks so tightly that it is difficult to remove it without tearing the animal.

 Distinctive features of anemones are the pharynx and sheets of tissue called mesenteries formed by in-foldings of body wall around the mouth. Anemones feed on small fish and invertebrates.

2. <u>Examine the preserved specimen of coral</u>. What you are observing is the protective limestone skeleton secreted by the small fragile polyps that once lived within it. Note the small holes in the coral. The living polyps extend through these holes for feeding. The polyps form huge colonies consisting of millions of individuals. New individuals build their skeletons on the skeletons of dead ones. Over a great many years huge undersea ledges, coral reefs, are built up.

3. Notice the preserved specimen of the colonial sea pen, another example of an anthozoan. These are "soft corals." The skeleton is proteinaceous rather than calcareous.

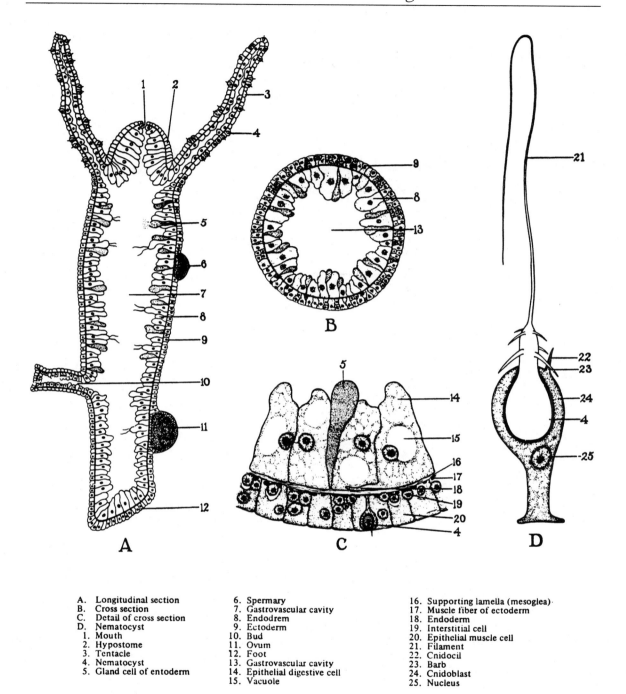

A. Longitudinal section
B. Cross section
C. Detail of cross section
D. Nematocyst
1. Mouth
2. Hypostome
3. Tentacle
4. Nematocyst
5. Gland cell of entoderm

6. Spermary
7. Gastrovascular cavity
8. Endodrem
9. Ectoderm
10. Bud
11. Ovum
12. Foot
13. Gastrovascular cavity
14. Epithelial digestive cell
15. Vacuole

16. Supporting lamella (mesoglea)
17. Muscle fiber of ectoderm
18. Endoderm
19. Interstitial cell
20. Epithelial muscle cell
21. Filament
22. Cnidocil
23. Barb
24. Cnidoblast
25. Nucleus

70764 - 01

Figure 4-4. Hydra Anatomy with Nematocyst Detail

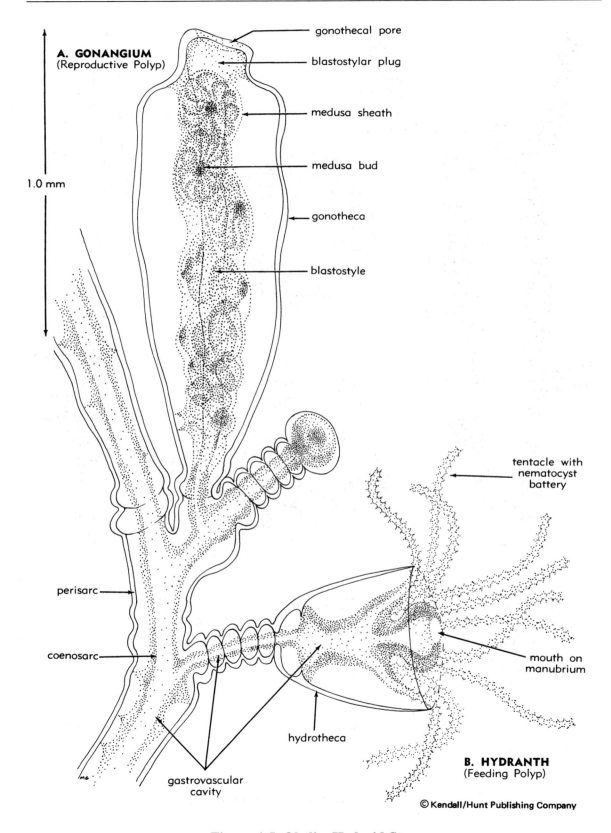

A. GONANGIUM
(Reproductive Polyp)

gonothecal pore

blastostylar plug

medusa sheath

medusa bud

gonotheca

blastostyle

1.0 mm

perisarc

coenosarc

tentacle with
nematocyst
battery

mouth on
manubrium

hydrotheca

gastrovascular
cavity

B. HYDRANTH
(Feeding Polyp)

© Kendall/Hunt Publishing Company

Figure 4-5. Obelia, Hydroid Stage

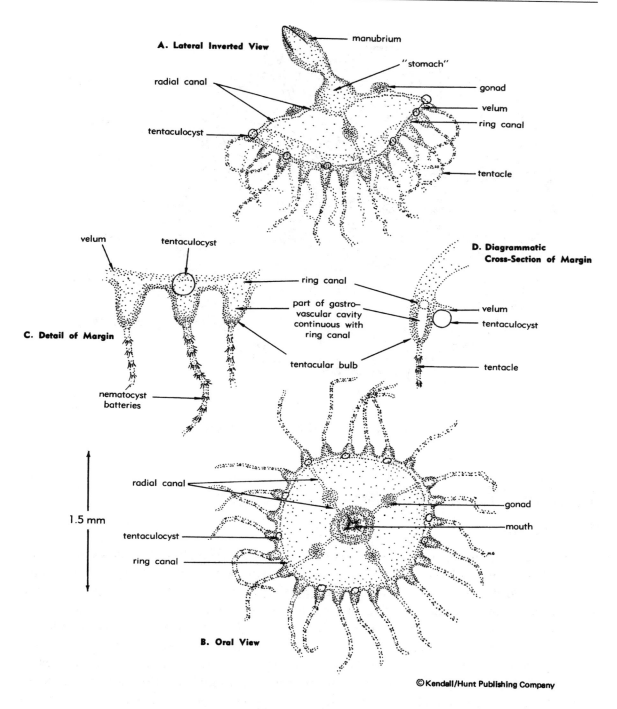

A. Lateral Inverted View

manubrium

"stomach"

radial canal

gonad

velum

ring canal

tentaculocyst

tentacle

velum

tentaculocyst

D. Diagrammatic
Cross-Section of Margin

ring canal

part of gastro—
vascular cavity
continuous with
ring canal

velum

tentaculocyst

C. Detail of Margin

tentacular bulb

tentacle

nematocyst
batteries

1.5 mm

radial canal

gonad

mouth

tentaculocyst

ring canal

B. Oral View

Figure 4-6. Obelia, Medusa Stage

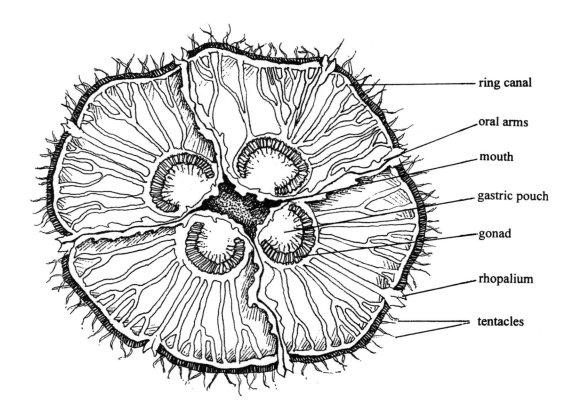

ring canal

oral arms

mouth

gastric pouch

gonad

rhopalium

tentacles

Figure 4-7. Aurelia, Whole Mount

Phylum Platyhelminthes

Background

The platyhelminthes are the flatworms and, as their name implies, they are dorsoventrally flattened. They are triploblastic, but do not have a coelom (body cavity). The space between the body wall and the gut is filled with a tissue (<u>mesenchyme</u>) derived from <u>mesoderm</u>, the third germ layer. As in the cnidarians, the digestive system, when present, is a gastrovascular cavity with one opening. Some of the characteristics shared with more advanced phyla are bilateral symmetry, well-developed organ systems, and well-developed reproductive organs. This is the first example of a dorsoventral differentiation with distinct anterior and posterior ends (<u>cephalization</u>). The sexes are usually united in one animal (hermaphroditic). Platyhelminthes can be divided into three major classes. Time limits our study to the planarian, <u>Dugesia</u> (class Turbellaria). Members of the class Turbellaria are free-living flatworms occurring in marine and freshwater environments. Some of these are considered parasitic. <u>Dugesia</u> is commonly studied as an example of a free-living form.

<u>Dugesia</u> (**Figures 4-10 and 4-11**) is a greatly flattened animal with clearcut <u>dorsal</u> and <u>ventral</u> sides. Since <u>Dugesia</u> moves exclusively in one direction (see if you can make one back up),

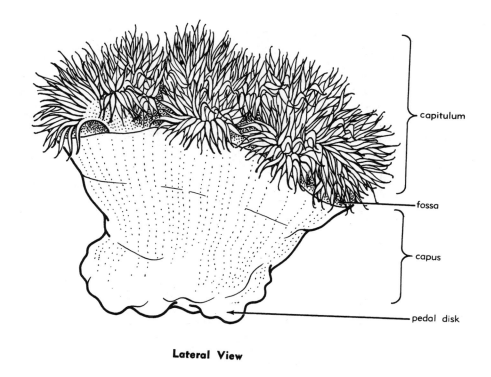

capitulum

fossa

capus

pedal disk

Lateral View

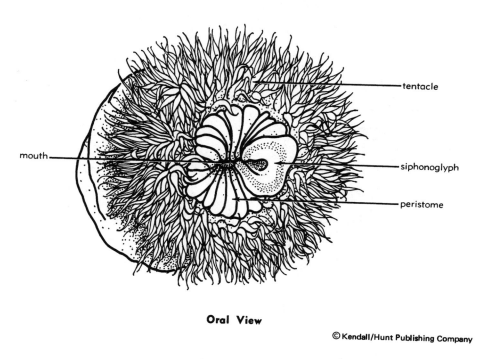

tentacle

mouth

siphonoglyph

peristome

Oral View

© Kendall/Hunt Publishing Company

FIgure 4-8. Metridium, External Views

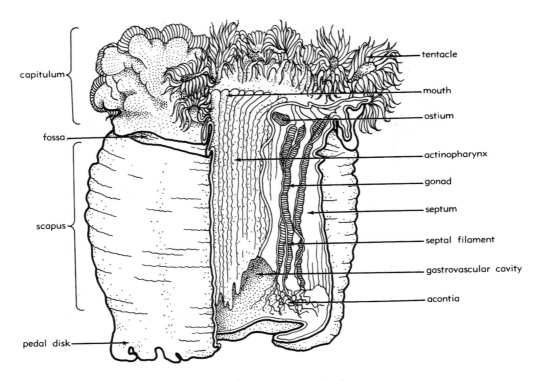

LATERAL VIEW (One Quarter Cut Away)

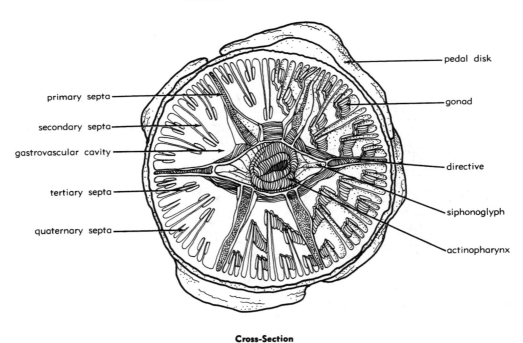

Cross-Section

Figure 4-9. Metridium, Sections

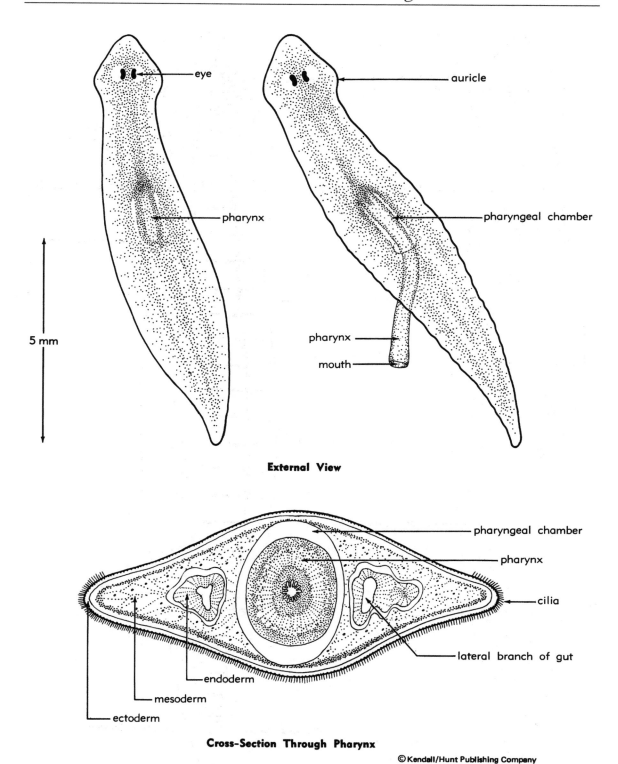

External View

Cross-Section Through Pharynx

Figure 4-10. Dugesia (Planaria)

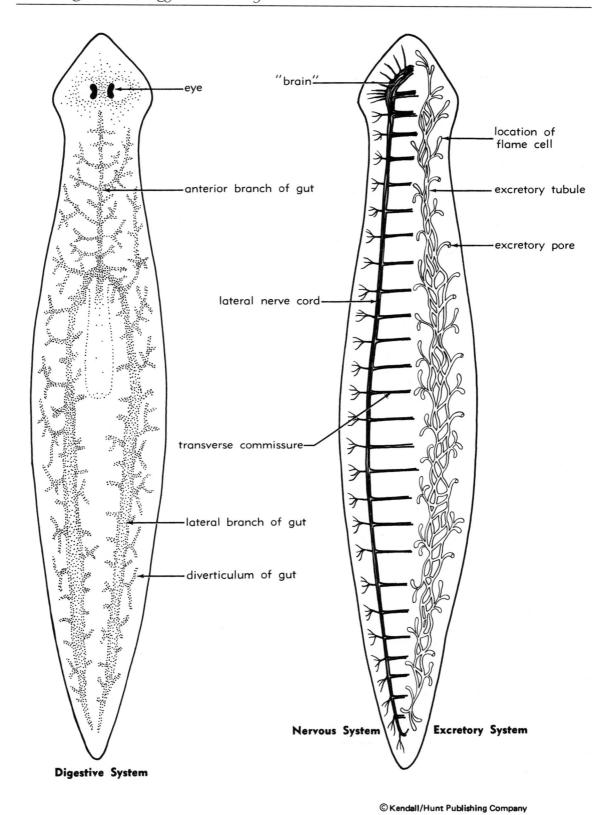

eye

anterior branch of gut

lateral branch of gut

diverticulum of gut

Digestive System

"brain"

location of flame cell

excretory tubule

excretory pore

lateral nerve cord

transverse commissure

Nervous System   **Excretory System**

Figure 4-11. Dugesia (Planaria) Whole Mount

it also has well-defined <u>anterior</u> and <u>posterior</u> ends. At the anterior end are located the sense organs, including the eyespots, which are sensitive to light, and the triangular <u>auricles</u> on each side of the head, which are sensitive to mechanical stimuli and are also chemosensory. The adaptive advantage of having sensory elements concentrated at the anterior end of the animal should be obvious. Other sensory cells of various types are scattered over the surface of <u>Dugesia</u>. If you can induce your animals to crawl over the surface of the water or on an inverted glass slide, you should be able to observe two pores on its ventral surface. The anterior pore is the opening to the <u>pharyngeal pouch</u>, from which the <u>pharynx</u> is protruded during feeding. The posterior pore is the genital opening, through which the sperm are exchanged during copulation.

Movement is accomplished primarily by ciliary propulsion over a layer of mucus secreted by adhesive glands. Muscle contractions in the form of rhythmic waves of movement sometimes cause locomotion. Try to distinguish the extent to which its movements result from ciliary action (smooth gliding) or from muscular activity (waves of contraction or changes in shape).

<u>Dugesia</u> hunts for its food. If the animals are hungry, they will soon respond to the presence of the food by actively moving until they have encountered it. When <u>Dugesia</u> has made contact with a fragment of food, it generally wraps itself around it. The pharynx protrudes through the mouth from a pouch on the ventral side. As <u>Dugesia</u> commences to swallow its victim, the pharynx can sometimes be observed through the dissecting microscope. Gas exchange takes place by diffusion through the body surface.

The Platyhelminths are the first metazoa you have studied that have a discrete excretory system. Its precise role is still open to some questions. Its main function may be osmoregulation. It is often absent in marine Turbellaria. The system consists of a series of tubules that open to the outside of the body via <u>nephridiopores</u>. The tubules end in ciliated <u>flame bulbs</u> (**Figure 4-12**). Cilia in the bulb maintain a flow of liquid along the tubule which is replaced by water and other substances that pass across the membranes into the flame bulb cell. It is easy to understand how this principle of a differentially permeable tubule membrane could evolve to the functioning chordate kidney.

Sexual reproduction takes place after gamete exchange with another individual. Self-fertilization is not possible. <u>Dugesia</u> can also reproduce asexually by fission. After the worm has grown to a specified length, its posterior end is pinched off from the anterior end and forms its own head and pharynx, etc., and thus becomes a complete adult. Asexual reproduction of this sort is frequently correlated with a high capacity for regenerating lost parts.

The nervous system is composed of two ventral longitudinal nerve cords that extend from the cerebral ganglia to the posterior end of the body. Cells in the eyespots are light sensitive. The auricles function in other sensory aspects such as smell and touch (**see Figure 4-10**).

Procedure

1. Obtain a watch glass containing a living planarian (<u>Dugesia</u>). Observe the external features of <u>Dugesia</u> as mentioned above in the background information.

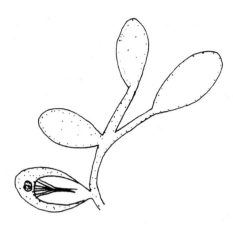

Figure 4-12. Flame Bulb Cells of <u>Planaria</u>

2. Recently, some students of mine did an experiment that involved giving a certain dosage of caffeine to planaria. To our surprise, a dosage of 0.25% caffeine caused planaria to display feeding behavior. This dosage was also lethal to the planaria within a few hours.

    I would like to continue this study in our lab exercise today. Each row of students will have a different dosage of caffeine. As a group you will treat a planaria with this dosage and monitor the effects over the lab period. Note the time in which the planaria was exposed to the caffeine. Also, note the feeding behavior at different concentrations of caffeine in your lab notebook. Don't forget to observe the feeding behavior and general health of other planaria treated with the different concentrations of caffeine indicated below. This experiment will have both a positive and negative control. Which treatment is the positive control? Which treatment is the negative control?

    Concentration of Caffeine:
    0%
    0.05%
    0.15%
    0.25%

3. **Planaria flame cells:** <u>One student/bench. Chop planaria with razor blade</u>. Prepare dry mounts and gently press on coverslip to squash the tissue. Examine under 40X for beating flagella. The cells are very small (about 810 mm in diameter). Look sharply!

4. **Internal Anatomy: Examine a prepared slide of the whole <u>Dugesia</u>. Then observe a slide of transverse sections of <u>Dugesia</u>.** Transverse sections from the anterior, pharyngeal, and posterior regions will be present on your slide. With the low power of the compound microscope observe a pharyngeal section, which is characterized by a conspicuous, circular <u>pharynx</u> surrounded by a cavity, the <u>pharyngeal pouch</u>. The opening through which the pharynx is protruded during feeding will probably not be visible in your section. On the ventral side of the animal, and on each side of the pharyngeal cavity, lie the nerve cords, which in section appear as poorly stained circular objects. Since they run the entire length of the animal, they should be visible in all sections. Now move

the slide so that you are viewing either an anterior or posterior section of the flatworm. Several cavities are visible in these sections. These are branches of the digestive cavity. Diverticula are lateral projections of the gastrovascular cavity that function to increase digestive surface area. They are continuous with the cavity in the center of the pharynx. A pair of nerve cords should also be visible in these sections. Scattered between the dorsal and lateral branches of the digestive cavity may be seen several solid masses of tissue, which are the testes. Fibrous structures that cross the body from the dorsal to the ventral side are the dorso-ventral muscle fibers, whose contractions serve to flatten the animal.

Turn to the high-power objective and observe the following structures, which are listed in order from the outside inward. The cilia are located on the surface of the animal, and their action is responsible for the gliding motion of the flatworms. Next is an inconspicuous layer of cells, to which the cilia are attached. Next is a muscle layer whose fibers are oriented in three distinct groups. There is an outer, circular layer, just beneath the epidermis; an inner, longitudinal layer; and dorso-ventral muscles that occur in strands. The structure of the mesoderm, which lies between the outer ciliated cells and the cells that line the digestive cavity, is extraordinarily complex.

Phylum Nematoda

Background

Most nematodes are small and soil-dwelling, but a great many are important parasites of plants and animals. Only a few parasites reach a size that is measurable in inches. These worms characteristically have round, tubelike bodies, and are, therefore, called roundworms and threadworms. They show bilateral symmetry and are nonmetameric (nonsegmented). They are triploblastic, having ectoderm (which makes up the dermal layer and cuticle covering the body), endoderm (the epithelium making up the whole of the digestive system), and mesoderm (which is found in the muscles and reproductive organs).

The roundworms show differentiation of anterior, posterior, dorsal, and ventral regions. These characteristics are similar to the flatworms. The phylogenetic advances made in the roundworm are a complete gut with a mouth and anus, separate sexes, and a body cavity. The body cavity is not a true coelom, as in the higher phyla. Since it is not completely lined with mesodermal derivatives, it is called a pseudocoel.

Procedure

Obtain a specimen of Anguillula, or vinegar eel. Notice the whip-like movement characteristic of this group.

Review Questions

1. Invertebrates are recorded in fossils over _____ mya.

2. Water enters the sponge through _____ and leaves via the _____ . The internal cavity is called the _____ .

3. Amoebocytes <u>do not</u>

   a. line the spongocoel

   b. secrete CaCO$_3$, SiO$_2$ or spongin

   c. carry food particles

   d. play a role in reproduction

4. <u>Obelia</u>

   Phylum?

   Class?

5. Jellyfish sense gravity using _____ .

6. Where on a cnidarians body are you most likely to find cnidocytes?

7. <u>Hydra</u> attaches itself to the substrate by its _____ .

8. Match:

   Planula larvae   asexual reproduction

   polyp   sexual reproduction

   medusa   habitat selection

9. Which class of cnidarians have only organisms of the polyp stage?

10. What are the simplest animals with bilateral symmetry and 3 tissue layers? Give phylum.

11. Identify the tissue layers in a jellyfish medusa.

12. Why are flatworms flat?

13. Nematodes move with a characteristic whipping motion. Why?

Answers

1. 600 mya

2. Ostia; osculum; spongocoel

3. Play a role in reproduction

4. Cnidaria; Hydrozoa

5. Statocysts

6. On the tentacles

7. Basal disc

8. Planular larvae — asexual reproduction
   polyp — sexual reproduction
   medusa — habitat selection

9. Anthozoans

10. Platyhelminthes

11. Endoderm, ectoderm

12. Oxygen can diffuse to all cells

13. Nematodes have longitudinal muscles, but no circular muscles.

References

Barnes, R.D. *Invertebrate Zoology*. W.B. Saunders Co. Philadelphia. 1974.

Brown, F.A., Jr. (ed). *Selected Invertebrate Types*. Wiley, NY. 1950.

Buchsbaum, R. *Animals Without Backbones*. U. of Chicago Press. 1976.

Engemann, J.G. and R.W. Hegner. *Invertebrate Zoology*. MacMillan Pub. Co. Inc. N.Y. 1981.

Hyman, L.H. *The Invertebrates*. McGraw Hill, N.Y. 1940.

Bio 204 Review Checklist

Lab 4—Lower Invertebrates   *indicates non-ID terminology

Phylum PORIFERA	*Grantia,* or *Scypha*	ostium spongocoel choanocyte	apopyle spicule feeding*
Phylum CNIDARIA Class Hydrozoa	*Hydra*	polyp only* tentacle basal disk ovary/spermary asexual bud gastrovascular cavity movement*	cnidoblast/cnidocyte cnidocil nematocyst hypostome gastrodermis mesoglea feeding*
	Obelia	perisarc gastrovascular cavity hydranth (feeding polyp) gonangium (reproductive polyp)	coenosarc medusa bud
	Physalia	medusa and polyp colony* nematocyst*	
Class Anthozoa	*Metridium*	polyp only* mouth gastrovascular cavity	pedal disk actinopharynx gonad
	Coralium	connecting sheet in colony*	
Class Scyphozoa	*Aurelia* (medusa)	tentacle rhopalium gastric pouch oral arm	gonad ring canal mouth polyp in larval forms
Phylum PLATYHELMINTHES Class Turbellaria	*Dugesia,* or *Planaria*	eyespot pharynx cilia flame cell nephridiopore* feeding*	auricle pharyngeal chamber mesoderm gastrovascular cavity movement*
Phylum NEMATODA	*Anguillula*	cuticle	movement*

Earthworm Dissection

Introduction

A careful and thorough dissection of the earthworm is required for this unit. Review the text, drawings and charts before attempting your dissection.

Phylum Annelida: Segmented Worms

This phylum contains the segmented worms, almost all of which are freeliving. Note that we have reached a branching point in the family tree of animal phyla. All organisms at this point and above have true coeloms. A true coelom is present for the first time in annelids. The annelid coelom is formed by a splitting of the embryonic mesoderm and is said to be <u>schizocoelous</u> (schizo- means split). This type of coelom will also be found among arthropods and mollusks. Note that these three groups are found on the same branch of the phylogenetic tree. Organisms on the other side of the branch point are also coelomates, but their coelom is formed in a different manner.

The coelom of annelids is compartmentalized into segments by <u>septa</u>. Coelomic fluid within the body cavity acts like a <u>hydrostatic skeleton</u> against which muscles work to change body shape. Like nematodes, annelids have a oneway digestive tract with a mouth, anus, and several specialized regions.

A dorsal mass of nerve cells forming a ganglion or "brain" and a ventral nerve cord provide a primitive nervous system. The circulatory system is closed, blood being confined to vessels.

Marine <u>polychaetes</u> (sand worms), <u>oligochaetes</u> (freshwater annelids and earthworms), and <u>leeches</u> are among the most common annelids.

External Morphology

The first four segments, or <u>somites</u>, make up the head region. The first segment is the <u>peristomium</u>. It bears the mouth which is overhung by a lobe, the <u>prostomium</u>. The anus is in the last segment. The clitellum secretes the egg capsules into which the eggs are laid and secretes a mucus for copulation. On the ventral side of the earthworm there are little bristles or setae used by the worm to prevent slipping, which are manipulated by small muscles at their bases.

There are many external openings other than the mouth and anus, as illustrated in **Figure 5-2.** The male pores on the ventral surface of somite 15 are conspicuous openings of the sperm ducts from which spermatozoa are discharged. Note the two long seminal grooves extending between the male pores and the clitellum. These guide the flow of spermatozoa during copulation. There are also smaller female pores on the ventral side of segment 14. These are the oviducts that discharge eggs. On segments 9–10 and 10–11 there are two pairs of small openings for the seminal receptacles. The nephridiopores are paired excretory openings located on the lateroventral surface of each segment except the first 3 and the last. A dorsal pore from the coelomic cavity is located at the anterior edge of the middorsal line on each segment from 8 or 9 to the last. Many earthworms eject a malodorous coelomic fluid through the dorsal pores in response to mechanical or chemical irritation or when subjected to extremes of heat or cold. The dorsal pores may also help regulate the turgidity of the animal. How would the loss of coelomic fluid affect the animal's escape mechanism (quick withdrawal into its burrow)?

Important Note:

Demonstrate your dissections for this exercise by drawings, sketches and a written account. Note the color of organs and other pertinent details. Structures often do not appear as they are drawn in textbook diagrams. Make certain your record reflects what you found. If you are uncertain about the identity of a structure, check with your instructor.

Procedure (Preserved Earthworm)

1. **External Structure (See Figure 5-1).**

   a. Examine a preserved earthworm. Use a dissecting microscope when necessary. Find pores and structures mentioned above in the background information, including the clitellum and peristomium.

   b. How many pairs of setae are on each segment, and where are they located? Use the hand lens or binocular microscope to determine this.

   c. Strip off a small piece of the iridescent noncellular cuticle. Float it out in water on a slide, cover, and examine with the microscope for pores through which mucus is discharged from gland cells in the epidermis. Reduce the microscope light if necessary.

2. **Internal Structure and Function (See Figures 5-3, 5-4, 5-5, and 5-6).**

   a. Cut the animal 1–2″ posterior to the clitellum. Save the anterior end for later dissection. From the posterior piece cut off 3 or 4 cross sections about 1–2 mm thick and mount with a small amount of water on a glass slide. Do not use a coverslip. Examine with a dissecting microscope. Locate all of the following components in the hand sections.

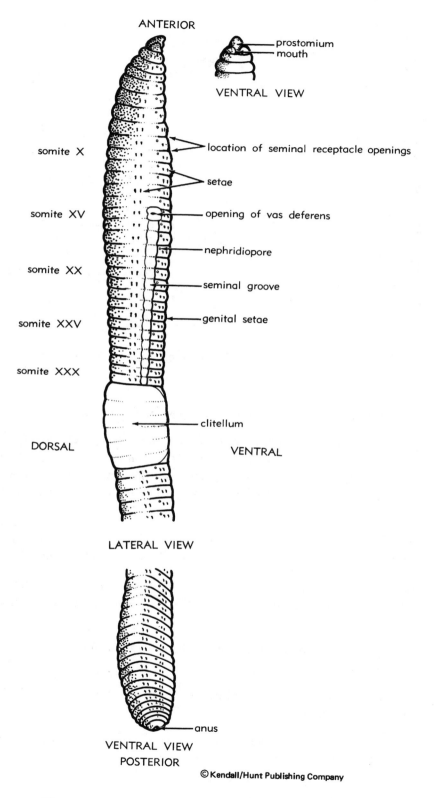

ANTERIOR

prostomium
mouth

VENTRAL VIEW

somite X

location of seminal receptacle openings

setae

somite XV

opening of vas deferens

nephridiopore

somite XX

seminal groove

somite XXV

genital setae

somite XXX

clitellum

DORSAL

VENTRAL

LATERAL VIEW

anus

VENTRAL VIEW
POSTERIOR

Figure 5-1. Lumbricus (Earthworm) External View

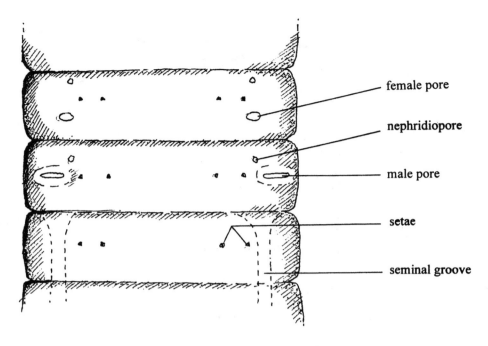

Figure 5-2. External Openings, Ventral View

b. Next, study the prepared slide c.s. and locate the same tissues.

c. <u>Intestine</u> The horseshoeshaped space at the center of the section is the cavity (lumen) of the intestine. The darkstaining tissue surrounding the lumen contains gut-epithelial cells, which absorb food molecules from the gut. Locate the dorsal fold in the intestine. It is called the (b) <u>typhlosole</u> and increases surface area for digestion and the uptake of the products of digestion.

d. <u>Muscle tissue</u> Two thin bands of muscle tissue surround the gut and are associated with the movement of food through the gut. These are of mesodermal origin.

e. <u>Chloragen tissue</u> Special digestive cells (lighter staining cells around the intestine) perform some of the functions of the liver in higher organisms. These cells are of mesodermal origin.

f. <u>Coelom</u> The large space bordering the digestive tract, lined by a sheet of mesoderm called the peritoneum.

g. <u>Nerve cord</u> Flattened, oval structure in the interior of the coelom. Notice that it is a ventral nerve cord.

h. <u>Blood vessels</u> One or two large blood vessels should be apparent next to the nerve cord. A fairly large dorsal vessel can be seen above the intestine. Anteriorly, this vessel gives rise to several stout lateral loops ("hearts") that encircle the intestine.

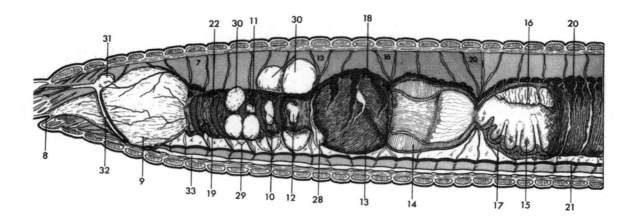

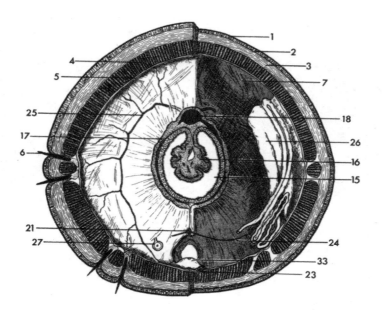

1. Cuticle
2. Epidermis
3. Circular muscles
4. Longitudinal muscles
5. Peritoneum
6. Seta
7. Coelom
8. Mouth
9. Pharynx
10. Esophagus
11. Esophageal pouch

12. Esophageal glands
13. Crop
14. Gizzard
15. Intestine
16. Typhlosole
17. Chlorogogen cells
18. Dorsal vessel
19. Lateral esophageal vessel
20. Intestinal vessel
21. Subintestinal vessel
22. Heart

23. Subneural vessel
24. Nephridial vessel
25. Parietal vessel
26. Nephridium
27. Nephrostome
28. Ovary
29. Seminal receptacles
30. Seminal vesicles
31. Suprapharyngeal ganglion
32. Nerve collar
33. Ventral nerve

 Carolina Biological Supply Company, Burlington, North Carolina 27215

Printed in U.S.A. © 1965 Carolina Biological Supply Company

Figure 5-3. Earthworm Anatomy—Dissection

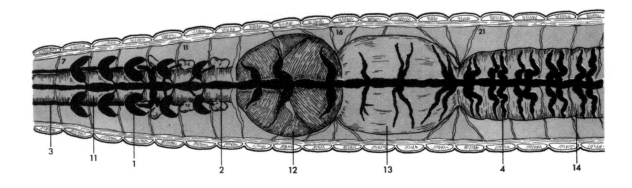

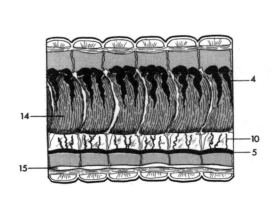

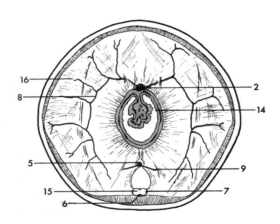

1. Heart	9. Nephridial vessel
2. Dorsal vessel	10. Ventral intestinal vessel
3. Lateral esophageal vessel	11. Esophagus
4. Intestinal vessel	12. Crop
5. Subintestinal vessel	13. Gizzard
6. Subneural vessel	14. Intestine
7. Lateral neural vessel	15. Nerve cord
8. Parietal vessel	16. Septum

Carolina Biological Supply Company, Burlington, North Carolina 27215

Printed in U.S.A. © 1965 Carolina Biological Supply Company

Figure 5-4. Earthworm Anatomy—Circulatory

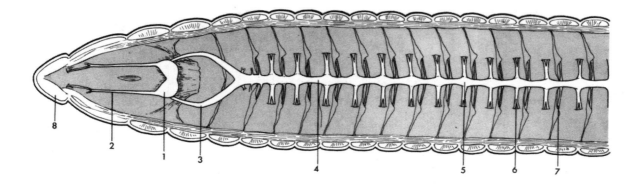

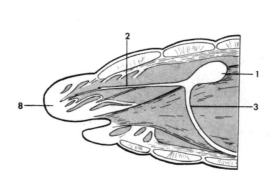

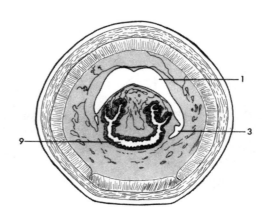

1. Suprapharyngeal ganglion
2. Anterior nerve
3. Nerve collar
4. Ventral nerve cord
5. Segmental ganglion

6. Segmental nerve
7. Septal nerve
8. Prostomium
9. Pharynx

Figure 5-5. Earthworm Anatomy—Nervous System

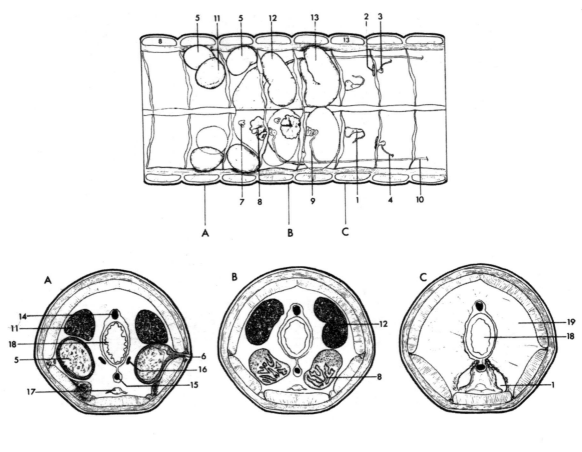

1. Ovary
2. Ovarian funnel
3. Ovarian sac
4. Oviduct
5. Seminal receptacle
6. Duct of seminal receptacle
7. Testis
8. Seminal funnel
9. Vas efferens
10. Vas deferens
11. Anterior seminal vesicle
12. Middle seminal vesicle
13. Posterior seminal vesicle
14. Dorsal blood vessel
15. Subintestinal blood vessel
16. Lateral esophageal vessel
17. Nerve cord
18. Esophagus
19. Septum

 Carolina Biological Supply Company, Burlington, North Carolina 27215

Printed in U.S.A. © 1972 Carolina Biological Supply Company

Reproduction of all or any part of this sheet without written permission from the copyright holder is unlawful.

Bioreview® Sheet
4374

Figure 5-6. Earthworm Anatomy—Reproductive System

i. <u>Nephridia</u> Inside each coelomic segment is a pair of excretory organs, the nephridia. Open ciliated funnels on the end of the nephridia filter the coelomic fluid and then carry it to the outside through excretory pores in the next most posterior segment. During the passage of fluid through these ducts, water can be resorbed and the contents modified as in the kidney tubule of vertebrates.

j. <u>Longitudinal</u> and j) <u>circular muscle bands</u> Two bands of muscular tissue are located on the inside of the body wall. Increase the magnification to observe the muscle arrangement in greater detail. Identify the longitudinal bundles and circular bands of muscle tissue. Because the earthworm is segmented its body is made up of a number of separate sections or compartments—the contraction of these muscles, acting in concert with the supporting hydrocoel present in each segment, can produce local movements that are quite different from the "lashing" activities of the vinegar eel. In general, contraction of the longitudinal muscles shortens (or bends) the worm while contraction of the circular muscles lengthens it. However, one part of the earthworm may be shortening or remain immobile while another part is lengthening. If live worms are available, observe their movements.

k. <u>Epidermis</u> Thin tissue surrounding the muscles. It secretes the 1) <u>cuticle</u> (body covering) and the slime trail upon which the earthworm glides. Respiration occurs by diffusion of gases through the epidermis.

l. <u>Setae</u> May be present in pairs as short stalks penetrating the integument. They are used to anchor one part of the worm's body while it moves another part forward. These structures can be felt as four rough rows along each side and each ventrolateral surface of an intact worm. See if you can feel the setae in the preserved specimen on demonstration.

3. Cut another piece of the worm from the **posterior** piece about 2.5 cm long and make a longitudinal section by cutting it in half dorsoventrally. Note how the <u>coelom</u> is divided into a series of compartments by <u>septa</u> that coincide with the external segmentation. In earthworms, these septa are incomplete ventrally where the nerve cord passes through, but these openings can be closed by sphincter muscles so that there is little movement of fluid between segments during locomotion.

4. Now holding the **anterior** portion of the worm dorsal side up, insert the scissors point into the cut posterior end just beside the dark middorsal line made by the <u>dorsal blood vessel</u>. Keep the lower blade up to prevent damaging the viscera. Cut close to one side of the vessel, extending the cut forward to about the fourth somite. Loosen the septa on each side to spread the body wall apart. Pin down each side of segment 15 (identified by the male pores). Then count the segments carefully and pin down segments 5, 10, 20, 30, and so on, thus creating definite landmarks with the pins. Slant the pins out at a wide angle from the body, occasionally moistening the specimen to avoid excess drying out.

5. Digestive System

Identify the mouth and locate the muscular <u>pharynx</u> attached to the body by dilator muscles for sucking action (the muscles, torn by the dissection, give the pharynx a hairy appearance); the slender <u>esophagus</u> in somites 6 to 13, which is hidden by the aortic arches and seminal vesicles; the large thin-walled crop (15, 16) for food storage; the muscular <u>gizzard</u> (17, 18) for food grinding; the <u>intestine</u> for digestion and absorption; and the <u>anus</u>. Two or three pairs of white calciferous glands are seen on each side of the esophagus; they are thought to secrete calcium carbonate to neutralize acid foods. Bright yellow <u>chloragen tissue</u>, covering the intestine and extending along the dorsal vessel, is conspicuous in the living earthworm. Chloragen cells are known to store glycogen and lipids, but probably have other functions as well, similar to those of the vertebrate liver.

6. Circulatory System

The well-developed blood system of the earthworm is a <u>closed system</u>; that is the blood flows in a continuous circuit of vessels rather than opening out into body spaces. Identify the <u>dorsal vessel</u> along the dorsal side of the digestive tract, five pairs of <u>aortic arches</u> encircling the esophagus in segments 7 to 11, and the vessel ventral to the digestive tract. The dorsal vessel is the chief pumping organ and the arches maintain a steady flow of blood into the ventral vessel. Lift up the intestine at the cut posterior end of the worm and see the ventral vessel. Now lift up the white nerve cord in the ventral wall and use the hand lens to see the <u>subneural vessel</u> clinging to its lower surface and the pair of <u>lateroneural vessels</u>, one on each side of the nerve cord.

7. Excretory System

A pair of tubular <u>nephridia</u> lie in each somite except the first three and the last. Each nephridium has a ciliated funnel-shaped nephrostome in the segment anterior to it. The <u>nephrostome</u> draws wastes from one somite into the ciliated tubular portion of the nephridium in the next somite, which empties the waste to the outside through a <u>nephridiopore</u> near the ventral setae. Use a dissection microscope to study the nephridia. They are largest in the region just posterior to the clitellum.

8. Nervous System

To examine the nervous system, do the following. Very carefully extend the dorsal incision to the first somite. Find the small pair of white <u>cerebral ganglia</u> (the brain), lying on the anterior end of the pharynx and partially hidden by dilator muscles; the small white nerves from the ganglia to the prostomium; a pair of <u>circumpharyngeal connectives</u>, extending from the ganglia and encircling the pharynx to reach the <u>subpharyngeal ganglia</u> under the pharynx; and a <u>ventral nerve cord</u>, extending posteriorly from the subpharyngeal ganglia for the entire length of the animal. Remove or lay aside the digestive tract and examine the nerve cord with a hand lens to see in each body segment a slightly enlarged <u>ganglion</u> and <u>lateral nerves</u>.

9. Reproductive System

The earthworm is monoecious; it has both male and female organs in the same individual, but cross fertilization occurs during copulation. First, consider the male organs. Three pairs of seminal vesicles (sperm sacs in which spermatozoa mature and are stored before copulation) are attached in somites 9, 11 and 12; they lie close to the esophagus. Two pairs of small branched testes are housed in special reservoirs in the seminal vesicles, and two small sperm ducts connect the testes with the male pores in somite 15; but both testes and ducts are too small to be found easily. The female organs are also small. Two pairs of small round seminal receptacles, easily seen in somites 9 and 10, store spermatozoa after copulation. You will probably not find the paired ovaries that lie ventral to the third pair of seminal vesicles or the paired oviducts with ciliated funnels that carry eggs to the female pores in the next segment. Note: you will obtain a clearer view of these systems by cutting out the digestive tract.

Dissect open the anterior seminal vesicle at its dorsal midregion. Place some of the contents on a slide and examine with a compound microscope. Compare with the prepared slide on demo. Spermatogonia and spermatocytes in various stages of development will be found (**Figure 5-7**). These preparations also commonly contain <u>Monocystis lumbrici</u>, a parasitic protozoan.

10. See other slides on demo in the lab for special histological features of the earthworm.

Figure 5-7. Morulae of Immature Sperm. Found in Seminal Vesicle

Procedures: Normal and Anesthetized Earthworm

This portion of the laboratory will be demonstrated to you by your TA.

1. Anesthetize the earthworm as directed by your TA. Work in pairs to conserve living material. Sharp instruments and fine forceps will be needed for some of these manipulations. Be certain of the earthworm anatomy before proceeding. Open middorsally by cutting carefully through the body wall with a razor blade or a new scalpel blade. Pin the animal as done for the preserved worm, but keep it moist with isotonic saline solution applied with a pipette. **Do not use water.**

2. Examine the peristaltic movements of the dorsal vessels and hearts. Note also peristaltic contraction of the gizzard.

3. Remove small pieces of the seminal vesicles with forceps. Gently tease them apart on a slide in saline solution. Examine for the fairly large trophozoite (200×70 μm) Monocystis.

4. Demonstration of nephridium. A method of demonstrating the nephridium, especially the difficult nephrostome, is to anesthetize a living earthworm. With a tuberculin syringe, inject a methylene blue solution into the body cavity. Wait 5 minutes. Make an incision in its middorsal region behind the clitellum, cut away the septa, and remove the intestine, being careful not to disturb the nephridia. Examine with the dissecting microscope. Small bulb-like structures will stain light blue. Why? These are the nephrostomes. With fine forceps remove one or two of these structures and prepare a wet mount. Study under 10 and 40X. Ciliary action will be readily apparent.

5. Remove small pieces of the seminal receptacles and tease the tissue apart in saline solution. Large numbers of spermatozoa may be present.

Lab Checklist

Kingdom: Animalia

Annelids (Oligochaeta, Earthworm)

General:
- feeding*
- movement*
- coelomate*
- hermaphroditic (monoecious)*

External:
- cuticle with pores
- somites/segments
- prostomium
- peristomium
- mouth
- clitellum
- male pore
- female pore
- nephridiopore
- seminal groove
- dorsal pore
- pores in cuticle
- seta (-e)

Digestive System:
- pharynx
- esophagus
- crop
- gizzard
- intestine
- typhlosole
- anus

Reproductive System:
- ovary
- oviduct
- seminal receptacle
- seminal vesicle
- morula
- sperm/spermatozoa

Circulatory System:
- closed system*
- dorsal blood vessel
- subneural blood vessel
- aortic arches ("hearts")

Excretory System:
- nephridia
- nephrostome

Nervous System:
- ventral nerve cord
- cerebral ganglia
- circumpharyngeal connectives
- subpharyngeal ganglia

Other:
- septum (-a)
- longitudinal muscles
- circular muscles
- chloragen tissue
- coelom
- peristalsis

References

Dales, R.P.: Annelids. London, Hutchingson University Library, 1963. A brief general account of the annelids.

Edwards, C.A., and J.R. Lofty: Biology of Earthworms. 2nd ed. London, Chapman & Hall, Ltd., 1977. This short volume covers the structure, physiology and ecology of earthworms.

Mill, P.J. (Ed.): Physiology of Annelids. London, Academic Press, 1978.

Phylum: Mollusca
Dissection of the Clam and Squid

Introduction

Mollusca represents a wide variety of species including snails, clams and squids. It is the second largest phyla of animals, containing more than 110,000 species. Most individuals are marine but large numbers are found in fresh water and some snails and slugs are terrestrial. Mollusca are the most recognized group of invertebrates. Mother of pearl inlay, shells and pearls—all products of mollusks—are all decorative items found in almost every home in North America. Shells not only provide tourists with a keepsake from the beach, but they have given evolutionary biologists a record of past invertebrate life forms. Shells, embedded in sediment on ocean bottoms, fix the approximate time of molluscan appearance on earth at about 600 million years ago.

Characteristics

All Mollusca have the same body plan, but the organization in each group may be quite different. There are three distinct features; the headfoot, the mantle and the visceral mass. Most Mollusca have an external shell, while slugs have no shell at all. The head-foot varies widely among individuals. It can be modified into long tentacles (octopus), a well-defined head (snails) or almost completely lacking (bivalves). Except in bivalves, all mollusca have a feeding organ at the head-foot region known as the radula. It is a rasping organ used to "saw away" at food stuff. The visceral mass of mollusca contains most of the internal organs except the gills (ctenidia). The gills are an extension of the mantle which surround the visceral mass. The outside portion of the mantle secretes the shell.

There are seven classes of Mollusca but over 98% belong to the classes Gastropoda, Bivalvia and Cephalopoda. A brief review of these three classes is given below. The focus of the lab will be a dissection of the clam and squid.

Class Gastropoda

Gastropoda contains over 80,000 species including snails and slugs. While most gastropods live in marine or brackish environments, many species are terrestrial. In fact, the gastropods are second only to arthropods in number of terrestrial species. Gastropods are ecologically important as food sources, agricultural pests and intermediates of some diseases. Escargot and some whelks are considered delicacies. Land snails and slugs account for the destruction of millions of dollars worth of crops each year due to their ravenous eating. However, a predacious species of snail can be used to destroy herbivorous gastropods. In some areas, snails are dangerous to humans as well as plants. In Asia, in particular, some parasitic nematode life cycles depend on snails as an intermediate before infecting human hosts.

Characteristics

Gastropods have a well-developed head-foot. Very muscular, the head-foot has developed into a distinct head region and a locomotive region. Protruding from the head are two antennae with sensory organs on each. They have a well-developed radula which grinds food and draws particles into the visceral cavity. Most species have a shell which takes many different forms. However, slugs do not retain their shells.

The land snail is a good example of this class. They feed on plant matter or algae. Large numbers of these snails invade vegetable farms, particularly in California where they destroy considerable amounts of lettuce and tomato plants.

Class Bivalvia

The clam will serve as our representative bivalve today. The body of animals in this class is laterally compressed and covered by a shell. The shell is formed by secretions from the two-lobed mantle. Inside the mantle is the body of the animal comprised of head, foot and viscera. Bivalvia includes clams, mussels, oysters and scallops. Bivalves are nearly everywhere in one form or other. There are about 13,000 marine species and more than 2,000 species in fresh water. We tend to overlook the freshwater ones. They are usually smaller than their marine cousins, but check for them as residents of sandy bottoms of most any stream or lake.

Most bivalves are filter feeders, and do not have a radula. They feed by beating their many cilia, creating a current of water which sweeps food particles to the mouth. The head-foot region is reduced to a burrowing foot. Sensory organs are not on the foot but at the edges of the mantle. Scallops have well-developed "eyes" that can distinguish light changes. The visceral mass is between a layer of mantle and two hard shells. These shells, secreted by the mantle, have a calcium carbonate matrix which makes them hard.

Class Cephalopoda

Cephalopods were once very dominant, but now contain only about 500 species. Very well known are the octopuses, squids and the pearly nautilus. Members of Cephalopoda are among the largest and most intelligent invertebrates. While the nautilus has retained its shell, the squid and cuttlefish have internalized shells, and the octopus has lost its shell altogether. Cephalopods have a well-developed brain and nervous system. Their eyes are very large (for invertebrates) and very accurate. An octopus can distinguish colors, while the squid's eyes are thought to be able to discern more details and depth than vertebrate eyes. Considered a delicacy in some parts of the world, cephalopods are used for food as well as for bait in fishing.

The squid is the largest known invertebrate. Although rarely seen, it is estimated that some squid reach the size of 21 m. An eye of a squid washed ashore in Australia was measured at 30 cm across (the largest eye of all animals) and sucker marks on sperm whales have suggested squid grow as large as 30 m! It is very probable that the legends of giant sea monsters were actually giant squid on the surface of the sea.

The Clam

Background

The adult life of the clam is spent burrowed head-down and posterior end up at or near the surface. The head of a clam has little else but a mouth flanked by feeding appendages, the palps. But, the clam is well-adapted to its lifestyle. It is a superb filter-feeder staying at home and slowly drawing great quantities of water over its perforated and ciliated gills. The system serves as a sieve, sorting out microscopic plants, animals, bacteria and other organic debris. From the ciliated gills to the palps and mouth, the food is sorted from other matter. At the posterior "neck" of the clam, a current of water is sustained passing in and then out through incurrent and excurrent siphons.

The clam burrows by extending its foot. Blood sinuses then fill to enlarge and anchor the organ while powerful retractor muscles draw the animal forward. Marine clams are harvested by digging them up at low tide. Some clams can actually move away through the mud faster than they can be excavated.

Most bivalves reproduce by separate sexes. Gametes are released through the excurrent siphon and fertilization occurs in the surrounding water. Some clams, including most freshwater species, brood their young on the female gills. There follows a sequence of larval stages (glochidia) before the clam settles down.

Procedures

1. Examine demo slides that review developmental stages of clam structure.

2. Orient the clam dorsal up with the shell umbo and anterior end to the left, as shown in **Figure 6-1.** The shells are attached by a hinge ligament on the dorsal side. Note the ventral side is open for extension of the foot. There are concentric growth lines around the umbo, with the youngest part at the edge.

3. **Opening the clam.** Locate the large anterior and posterior adductor muscles. These muscles and neighboring smaller ones must be cut to open the clam. With instructions from the TA, first cut the anterior muscle. The posterior muscle must be cut carefully to avoid damage to the pericardial sac. Cut from the dorsal side down and away from the heart

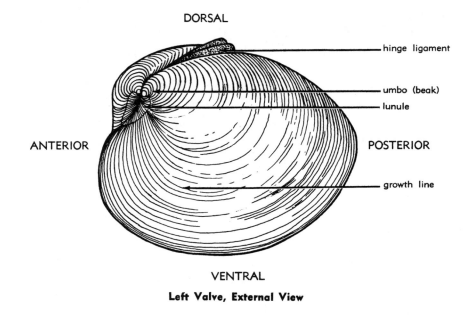

DORSAL

hinge ligament

umbo (beak)

lunule

ANTERIOR

POSTERIOR

growth line

VENTRAL

Left Valve, External View

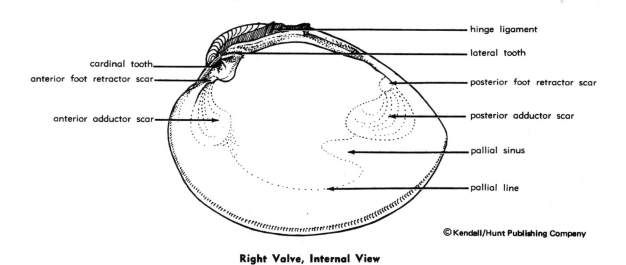

hinge ligament

lateral tooth

cardinal tooth

anterior foot retractor scar

posterior foot retractor scar

anterior adductor scar

posterior adductor scar

pallial sinus

pallial line

© Kendall/Hunt Publishing Company

Right Valve, Internal View

Figure 6-1. Clam External and Internal Views

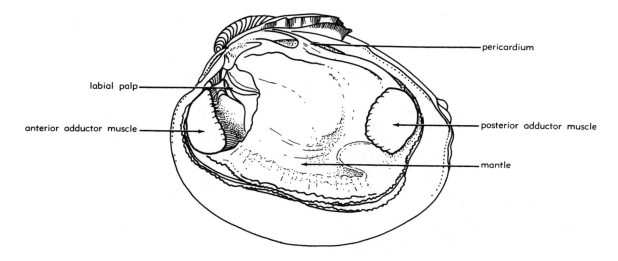

Left Valve Removed, Mantle Intact

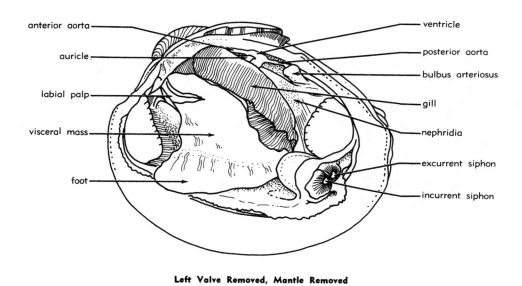

Left Valve Removed, Mantle Removed

Figure 6-2. Clam Anatomy

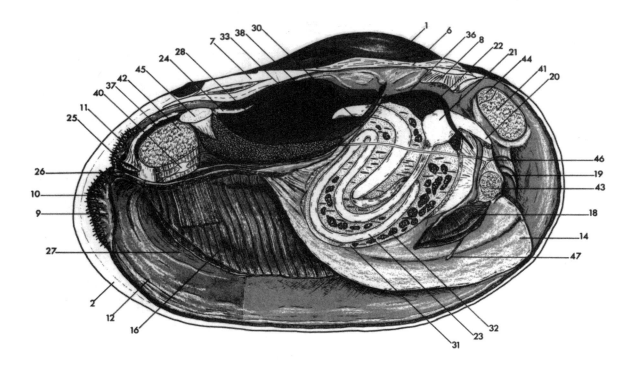

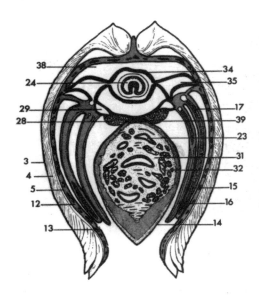

1. Umbo
2. Shell
3. Periostracum
4. Prismatic layer
5. Nacreous layer
6. Hinge ligament
7. Hinge
8. Hinge tooth
9. Incurrent siphon
10. Tentacles
11. Excurrent siphon
12. Mantle
13. Pallial line
14. Foot
15. Gill, outer lamina
16. Gill, inner lamina
17. Water tube
18. Labial palps
19. Mouth
20. Esophagus
21. Stomach
22. Liver
23. Intestine
24. Rectum

25. Anus
26. Cloacal cavity
27. Mantle cavity
28. Kidney
29. Bladder
30. Urinary opening
31. Visceral mass
32. Gonad
33. Heart
34. Ventricle
35. Auricle
36. Anterior aorta
37. Posterior aorta
38. Pericardium
39. Vena cava
40. Visceral ganglion
41. Anterior adductor
42. Posterior adductor
43. Foot protractor
44. Anterior foot retractor
45. Posterior foot retractor
46. Cerebral ganglion
47. Pedal ganglion

Carolina Biological Supply Company, Burlington, North Carolina 27215

Printed in U.S.A.© 1965 Carolina Biological Supply Company

Bioreview® Sheet
4380

Figure 6-3. Clam Anatomy

region. The clam can now be opened cutting away one side of the mantle to expose the body of the clam. Examine the inner part of the shell for teeth that fit tongue-and-groove fashion to keep the valves in position. Note again the <u>mantle</u> or <u>pallium</u>. It is a unique organ of all Mollusca. The mantle is a lateral outgrowth from the dorsal aspect of the visceral mass surrounding the body of the clam like an umbrella. Note its attachment to the shell along the <u>pallial line</u>. The free, distal edge of the mantle bears an outer secretory lobe that produces the shell. The middle sensory lobe and inner muscular lobe can move to regulate water flow.

4. Consult the figures and the instructor to locate the structures of the clam. Note the ventral <u>incurrent</u> and dorsal <u>excurrent siphons</u>. Locate the <u>labial palps</u> which surround the mouth. Other structures are the <u>gills</u> which normally are suspended in the mantle cavity, and the <u>foot</u>. Dorsal to the gills is the <u>pericardial sac</u>, which encloses the <u>heart</u>. Open the sac. The heart is 3 chambered. It has two <u>lateral auricles</u> and one <u>ventricle</u>. The <u>bulbus arteriosus</u> is a conspicuous enlargement on the posterior end of the ventricle. It may function to protect ventricular function from a back flow of blood caused by the sudden constriction of the posterior end of the clam. A sagittal section through the visceral mass should reveal the orangey <u>gonad tissue</u> and the greenish <u>digestive gland</u>, and may show <u>stomach</u> and <u>intestine</u>.

The Squid

Background

Unlike the sedentary clam, the squid has a fast-moving, predatory life. It captures many forms of marine life, but in turn, it is on the menu of larger fish and even sperm whales. The squid is probably the most highly evolved invertebrate. A muscular funnel provides the animal with fluid jet propulsion. Fins aid in abrupt changes in direction. An ink sac of special design can release a cloud of inky fluid to obscure a hasty retreat. Ten sucker-bearing arms surround the head. Two arms that are longer than the others have suckers only at the tips to shoot forward and grasp the prey. All of the arms then hold the unfortunate meal as two powerful jaws assist in introducing a toxin that further immobilizes the animal.

The nervous system of the squid is very well developed. The brain has several prominent ganglia connected together and two large, well-developed eyes.

Reproduction in squids involves separate sexes. The male has a specialized arm, called the hectocotylus, for transferring packets of sperm contained in spermatophores. Conjugal rights are apparently determined by male size, aggressiveness, and vigilance of the female. During mating the male wraps his arms around the female and deposits spermatophores directly on the forehead of the female or, on more rarely observed cases, inserts his hectocotylus inside the female mantle cavity, holding the spermatophores until sperm are released. After mating, female eggs are deposited in gelatinous egg masses.

The prey of squid consists of various crustaceans, fish, gastropods, bivalves, smaller squid and other invertebrates. After capture, the animal is crushed by the powerful beak-like jaws. The posterior salivary glands produce a toxin (tyramine) which is injected into the prey. The barbed radula functions like a conveyor belt, passing chucks of meat back to the esophagus which opens into a muscular stomach. The stomach in turn opens into a large caecum which fills much of the dorsal body cavity. A complex digestive gland (liver/pancreas) lying beside the esophagus and ending at the stomach produces enzymes that enter the digestive tract through a duct that opens at the juncture of the stomach and caecum. Much of the digestion occurs in the caecum. It is a single organ with two parts. Near the stomach it is spirally coiled. The ducts from the digestive gland are there, as are the openings from the stomach and intestine. On the inner walls there is a system of radiating pleats with grooves between them which center upon the opening of the intestine. Ciliary currents on the pleats and in the grooves collect particulate material and bind it with mucus. This waste is then moved along into the intestine. The second part of the caecum is a large expandable sac with smooth walls bearing a ciliated epithelium. The walls of the cae-cal sac are muscular and capable of contraction. The ciliary currents on the walls serve to cir-culate the fluid contents.

Details of the digestive process are reviewed by Purchon (1977). Some of the information is given here to illustrate the highly specialized and evolved digestive process of the squid. Food is passed through the esophagus by peristalsis. The interior of the esophagus is lined with cuticu-lar epithelium to protect it from bones, shells and the like. Similarly, the stomach is lined also by a cuticular sheath which also protects it as the food is churned by powerful stomach muscles. Prior to the meal, a supply of pancreatic enzymes accumulate in the caecum and are transferred to the stomach when food arrives. Food remains in the stomach for $1\frac{1}{2}$ to 2 hrs. Fluid contents, together with small particles, then pass through a sphincter into the caecum. Also at this time, secretions from the liver portion of the digestive gland which have accumulated in the lower por-tion of the ducts pass through relaxed sphincter muscles of the main hepatic ducts into the cae-cum. With these secretions, digestive is completed in 3–4 hrs. During this period the sphincter between the stomach and the caecum is closed. The sphincter muscles at the head of the hepatic ducts are also closed, preventing the release of more hepatic secretions, while pancreatic secre-tions can still pass down the lower parts of the ducts and into the stomach. In this way, stomach digestion of recently swallowed food can proceed while digestion of an earlier meal is completed in the caecum. It is a timely, coordinated, efficient process, all presumably under the control of the sympathetic nervous system of which the splanchnic ganglion is nearby. Finally, we can add that the walls of the intestine and rectum are protected by a liberal production of mucus added to the flow of waste material.

Procedure

1. External anatomy

a. Orient the squid so that the head and tentacles are toward you and the underside of the squid where the funnel is located is facing down (**Figures 6-4 and 6-5**). The head and tentacles represent the ventral surface of the body. Opposite from the head is the dor-sal end. The "funnel" or siphon surface is posterior and the anterior surface is fac-ing up. The body of the squid is much like a tube with the head inserted at one end

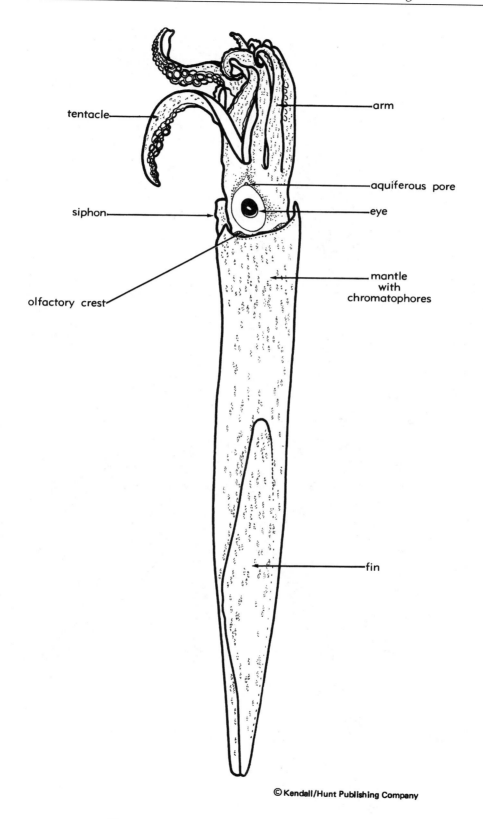

tentacle

arm

aquiferous pore

siphon

eye

olfactory crest

mantle
with
chromatophores

fin

Figure 6-4. Squid, External Lateral View

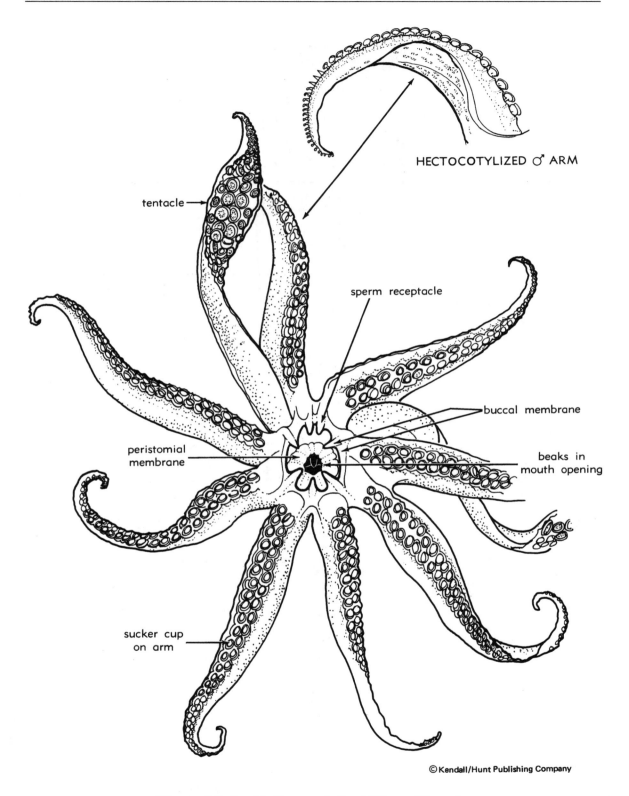

HECTOCOTYLIZED ♂ ARM

tentacle

sperm receptacle

buccal membrane

beaks in
mouth opening

peristomial
membrane

sucker cup
on arm

Figure 6-5. Squid, External, Oral View of Female

and flattened at the other. The outer tubular structure is the muscular mantle. Consider how different this structure is in form and function, although it is embryologically identical to the clam.

b. Note the pigmented surface of the body. The spots are chromatophores which are connected to muscle fibers that enlarge or retract the size of the pigment region depending on environmental conditions or predator/prey interactions.

c. On the head, locate the four pairs of arms with suckers and the two elongated tentacles. The fourth left arm of the squid is the hectocotylized appendage modified for the transfer of spermatophore. In our specimens (Loligo spp) only a slight modification in the localized absence of a few suckers may be observed. In other squids, suckers are completely absent and the surface of the arm has a sticky surface membrane. Take the time to remove a sucker and examine it under the **dissecting scope**. The sucker has a rounded cup attached by a stalk or pedicle to the arm. Note the chitinous ring of teeth around the margin of the sucker and the small piston-like structure at the base of the cup (**see Figure 6-6, #3**).

d. The mouth with two beaklike jaws is inside the arms. It is surrounded by the peristomial membrane. Outside of this is the buccal membrane with seven projections, each with suckers on the inner surface.

e. Locate the funnel (siphon) on the posterior (underside) of the animal. The siphon of the squid is not homologous to the clam siphon. The clam siphon is part of the mantle. Here, along with the arms and tentacles, it is a modification of the foot.

f. Also on the head, note the eyes capable of forming images. Ventral to each eye is a small aquiferous pore leading to a chamber that admits water to regulate the pressure within the eye chamber. An olfactory receptor known as the olfactory crest consists of a fold of tissue just dorsal to the eye.

2. **Internal anatomy**

a. Orient the squid with the siphon facing upward. Cut from the collar to the far end of the body being careful not to disturb the internal organs. Cut the mantle away on each side to expose the structures of the mantle cavity. At this time locate the stellate ganglion on the surface of the body cavity. Your TA will assist. The siphon should be cut lengthwise. Note the lip-like structures called funnel valves at the forward edge. These help regulate release of water from the mantle cavity.

b. Determine the sex of the squid before any organs are disturbed (see figures for clarity). Arrange for alternate male and female squid specimens at each lab bench so that the reproductive organs can be easily studied.

(1.) A female squid will have two oblong nidamental glands on top of the visceral mass. They secrete coverings for the egg capsules. The accessory nidamental glands are smaller, rounded organs just below the nidamental glands. They secrete an elastic membrane around each egg. Without disturbing the ink sac, carefully remove the nidamental gland. Find the oviducal gland which secretes the egg coverings, the long oviduct and the flared oviducal opening. At the dorsal end is a large mass of eggs contained in the ovary.

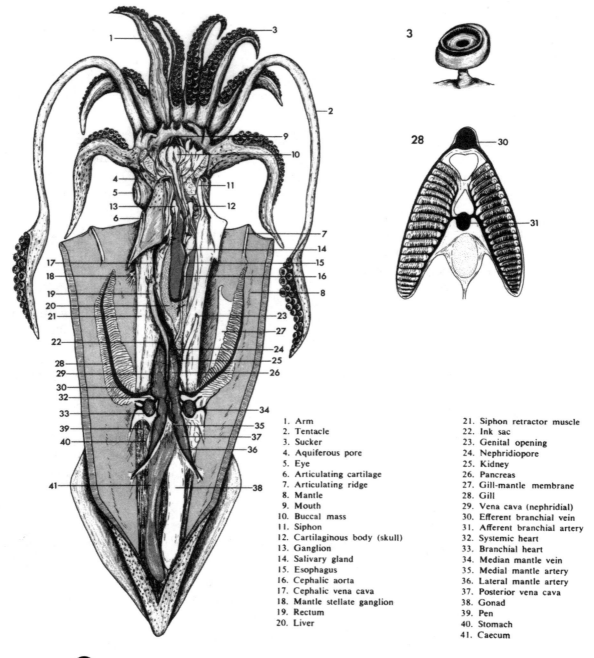

1. Arm
2. Tentacle
3. Sucker
4. Aquiferous pore
5. Eye
6. Articulating cartilage
7. Articulating ridge
8. Mantle
9. Mouth
10. Buccal mass
11. Siphon
12. Cartilaginous body (skull)
13. Ganglion
14. Salivary gland
15. Esophagus
16. Cephalic aorta
17. Cephalic vena cava
18. Mantle stellate ganglion
19. Rectum
20. Liver
21. Siphon retractor muscle
22. Ink sac
23. Genital opening
24. Nephridiopore
25. Kidney
26. Pancreas
27. Gill-mantle membrane
28. Gill
29. Vena cava (nephridial)
30. Efferent branchial vein
31. Afferent branchial artery
32. Systemic heart
33. Branchial heart
34. Median mantle vein
35. Medial mantle artery
36. Lateral mantle artery
37. Posterior vena cava
38. Gonad
39. Pen
40. Stomach
41. Caecum

Carolina Biological Supply Company, Burlington, North Carolina 27215
Printed in U.S.A. © 1967 Carolina Biological Supply Company
Reproduction of all or any part of this sheet without written permission from the copyright holder is unlawful.

Figure 6-6. Squid Anatomy—General

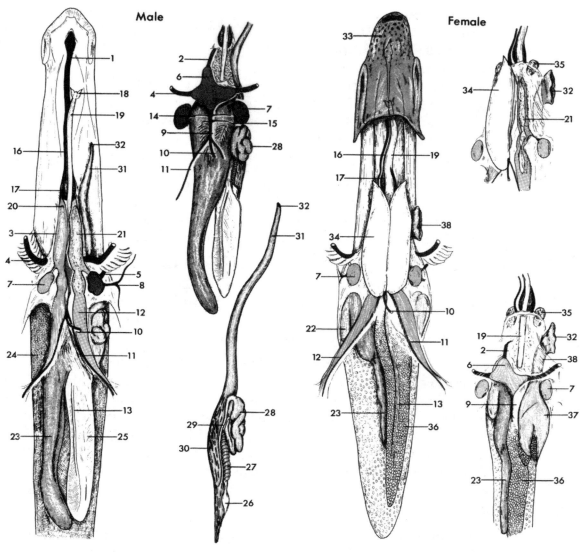

1. Cephalic vena cava
2. Cephalic aorta
3. Nephridial portion of vena cava
4. Efferent branchial vein
5. Afferent branchial artery
6. Systemic heart
7. Branchial heart
8. Median mantle vein
9. Posterior aorta
10. Medial mantle artery
11. Lateral mantle artery
12. Posterior vena cava
13. Genital artery and vein

14. Artery to branchial heart
15. Artery to gonoduct
16. Ink duct
17. Ink sac
18. Rectal papilla
19. Rectum
20. Nephridiopore
21. Kidney
22. Stomach
23. Caecum
24. Pen
25. Testis
26. Sperm bulb

27. Vas deferens
28. Spermatophoric gland
29. Spermatophoric duct
30. Spermatophoric sac
31. Penis
32. Genital opening
33. Siphon
34. Nidamental gland
35. Accessory nidamental gland
36. Ovary
37. Oviducal gland
38. Oviduct

Figure 6-7. Squid Anatomy—Circulatory & Reproductive

(2.) Consult the figures and your TA to locate the male organs. Locate the lobed testis. It is a flattened, fairly large organ in the dorsal portion of the cavity. Find the spindle-shaped spermatophoric sac. Next to it is a coiled vas deferens. The free end is enlarged and is called a sperm bulb. Sperm released from the testis are swept by ciliary action into the sperm bulb, through the vas deferens to the spermatophoric gland. Here the sperm are packaged into spermatophores. The spermatophores are then transferred to the spermatophoric sac (seminal vesicle). The spermatophores exit through the penis.

(3.) Examine a wet mount of a spermatophore under 10x (**Figure 6-8**). Dissect out the male reproductive structures. Examine the coiled vas deferens under the dissecting scope. The pearly white color is due to enclosed sperm. You may make a squash preparation and examine under 10x and 40x objectives. The testis won't show much and can be left in place. Avoid damaging the caecum.

c. Next, carefully cut away the membrane containing the visceral mass. Try not to tear any organs. Dorsal to the funnels locate the silvery ink sac under the rectum. It is filled with a melanin pigment. Remove the sac by carefully cutting it away from the rectum. Avoid puncturing the sac while it is attached to the specimen. You may wish to observe the ink dispersed in a beaker of water.

d. Note the paired gills which function to oxygenate the blood and the large blood vessel along the outer edge of the gill which connects to the systemic heart. Unlike the clam where blood circulates through open spaces in the body, the squid has a closed circulatory system. Blood flows through a continuous system of vessels with oxygen/CO_2 exchange occurring through capillaries in the gills. There are two pump mechanisms. Deoxygenated blood enters the branchial or gill hearts which push the blood out into the gills. The oxygenated blood from the gills enters a single systemic heart where it is pumped to the tissues. The two white lobes at the base of the gills are the branchial hearts.

e. Note the kidneys also near the base of the gills and extending ventrally. These are difficult to recognize. Be certain to locate them with the assistance of your TA. Also locate the paired nephridiopores just behind the ink sac. Just under the nephridia find the large white systemic heart with arteries extending to the body walls. The two white blood vessels dorsal to the branchial hearts are the vena cava (posterior).

f. Dissection of the visceral mass.

(1.) It is possible to carefully dissect out much of the digestive system. With help from your TA, locate the stomach. Study your animal from a side view. The objective will be to dissect out the stomach, the caecum, and a portion of the esophagus. Proceed carefully as the wall of the caecum is easily ruptured.

(2.) Locate the ventral portion (front) of the caecum. The previously mentioned region with ciliated grooves should be located. (Students who satisfactorily complete this aspect of the dissection will possibly be given special recognition and credit by their peers and the TA!)

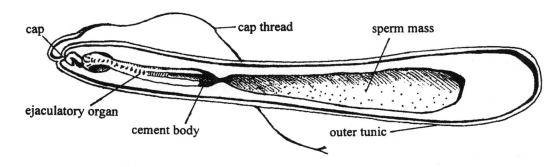

Figure 6-8. Spermatophore

(3.) Before discarding the squid body section, locate the <u>pen</u>. It is the shell of the squid represented by only a thin dorsal strip of horn-like material. The shell is totally absent in the octopus.

g. The final aspect of the dissection is to prepare a sagittal section (**Figure 6-9**) of the head of the squid. Again, your, TA will provide details. Locate the <u>beak</u>, <u>radula</u> (remove and examine under the dissecting scope), <u>buccal bulb</u>, <u>cephalic cartilage</u>, <u>cerebral ganglion</u>, and the <u>esophagus</u>.

This concludes our dissection of this remarkable animal. With more time there is much more we could do. Even so, our dissection serves to demonstrate the complexity of structure and function of some of the invertebrates.

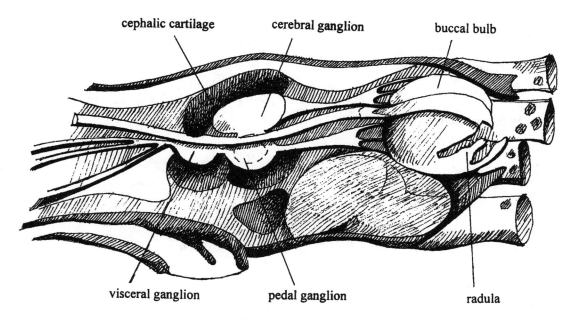

Figure 6-9. Squid Head, Sagittal Section

Clam—Checklist

Dorsal, ventral, posterior and anterior positions
Umbo
Hinge ligament
Shell growth lines
Teeth
Pallial line
Incurrent and excurrent siphon
Mantle
Foot
Gill
Labial palps
Anterior and posterior adductor muscle.
Parts of the digestive system: (to the extent possible) mouth, esophagus, stomach, digestive gland, intestine, rectum and anus
Gonad tissue
Circulatory system: pericardial sinus or cavity, pericardium, ventricles, auricle, bulbus arteriosus, anterior and posterior aorta

Squid—Checklist

External Anatomy:
    tentacles
    arms
    buccal membrane
    peristomial membrane
    chitinous beak
    eyes
    funnel
    chromatophore cells
    fins

Internal Anatomy:
    funnel valves
    gills
    ink sac
    rectum
    kidney
    nephridial pores
    branchial hearts
    systemic heart
    buccal bulb
    esophagus
    stomach
    intestine

rectum
caecum
cephalic cartilage
cerebral ganglion
pedal ganglion
stellate ganglion
visceral ganglion
radula, viewed under dissecting scope
pen

female:
nidamental glands
accessory nidamental glands
oviducal gland
oviduct
oviducal opening
ovary

male:
penis
spermatophoric gland
spermatophoric sac
spermatophore, 10X & 40X
testis

References

Mollusca

Gardiner, M. 1972. The Biol of Invertebrates. McGraw Hill.

Hyman, L.H. 1967. The Invertebrates: Mollusca (Vol. VI). McGraw Hill, N.Y.

Morton, J.E. 1958. Mollusca. Hutchinson, London.

Nison, M. and J.B. Messenger (Eds). 1977. The Biology of Cephalopods. Academic Press, London.

Purchon, R.D. 1977. The Biology of the Mollusca. 2nd ed. Pergamon Press, Oxford.

Wilbur, K.M. and C.M. Yonge. 1964. Physiology of Mollusca. Vol I. Academic Press, New York.

Phylum: Arthropoda

Introduction

Arthropoda comes from the Greek "arthro" meaning jointed and "poda" meaning feet. The "jointed feet" refers to the many jointed appendages of all arthropods. The arthropods contain the largest number and largest diversity of species in the animal kingdom. Indeed, three out of four animals in the world is an arthropod. In virtually every habitat an arthropod species can be found. Over a million species have been identified and some estimates go as high as two million unidentified species.

Insecta, a class of arthropods, is economically the most important group of animals. Almost every plant has an herbivorous insect species that feeds upon it. However, without insects many plants would not be pollinated each year. Outbreaks of many human diseases can be associated with insect populations. Insects serve as vectors in the transmission of such diseases as: Bubonic Plaque (flea), Rocky Mountain Spotted Fever (tick), and Malaria (mosquito). Other Insecta (lady bird beetles, mantis) and the class Arachnida (spiders, scorpions) can be used as biological controls to many agricultural pest problems. Crustaceans, another group of arthropods, also have economical importance. Large crustaceans (decapods) particularly lobsters, shrimp, and crabs serve as a major source of food in some areas of the world. Fresh water crustaceans, crayfish, are also important in the diet of many areas. A small crustacean, the barnacle, has been a problem for many years. Attaching to the bottom of boats and ships, they cause the loss of millions of dollars a year in fuel consumption.

Characteristics

All arthropods are segmented and bilaterally symmetrical. The body is typically divided into three regions: the head, the thorax, and the abdomen. They are covered with a chitinous exoskeleton which provides protection and attachment points for muscles. As the arthropod grows, it sheds its exoskeleton by molting. Many arthropods go through drastic changes in form when they molt (metamorphosis). In this case the larvae bears no resemblance to the adult (e.g., caterpillar and butterfly).

There are many similarities between arthropods and the annelids (earthworm) that form the basis of the suggestion that arthropods evolved from the annelids. Both groups are triploblastic, have a ventral nerve cord, paired and segmentally arranged appendages, segmentation, and a vascular system with a dorsal heart. Some differences between these two goups are that many arthropods (including the grasshopper) have replaced the coelom with a new body cavity called the hemocoel.

Class Crustacea

The name Crustacea (L., crusta, shell) is given for the hard outer exoskeleton of these animals. It is a familiar group to us and includes lobsters, crayfish, crabs, shrimp, copepods and waterfleas. The group is mostly marine, aquatic and free-living. But, there are many exceptions. Crustacea are an important food source for man. They are also an important link in the food chain for other animals. For example, the largest living animals, baleen whales, live on a diet of krill, a very small, but abundant shrimp-like creature.

Crustaceans show 3 characteristic arthropod features. The body is segmented, covered by a hard exoskeleton and the appendages are jointed. They are distinguished from other arthropods by 2 pairs of antennae. Typically also on the head is a pair of mandibles and 2 pairs of maxillae. This is followed by a pair of appendages on each body segment. The form for these appendages closely fits specialized and often different functions.

Class Insecta

Another example of an arthropod that we will observe today is the grasshopper. It is a member of the class Insecta. These organisms are terrestrial and have been largely successful due to their small size, the evolution of wings, and the ability to undergo metamorphosis. These three traits have allowed them to specialize to fill specific niches in their environment. The grasshopper is considered a typical representative of the Insecta because it is relatively unspecialized and has not greatly differentiated from the appearance of ancient insect groups. Therefore it is useful to fully understand grasshopper morphology before studying groups that are more highly modified.

Development of insects may involve metamorphosis from a juvenile to an adult stage. Insects that undergo a distinct metamorphosis can be divided into two groups. Hemimetabolous insects exhibit early developmental stages (**nymphs**) that resemble adults in most external features and holometabolous insects in which the young (larvae) do not resemble the adults at all (caterpillar to butterfly). The grasshopper is a hemimetabolous insect. Another term used for this type of development is **incomplete metamorphosis**.

The Crayfish

Background

The crayfish will be studied for basic features of an arthropod and of course, a crustacean. Crayfish have a worldwide distribution. However, there are about 350 species in the United

States, more than anywhere else in the world. The greatest density of species occurs in our southeast states, particularly along the Gulf Coast. Our live study specimens are from Louisiana. The world's largest museum collection of crayfish, totaling about 5 million individuals, is held by the Smithsonian Institution in Washington, D.C.

The crayfish is an interesting animal. It lives in the fresh water of ponds, lakes and streams. During daylight, they stay hidden under rocks or other protective places, then feed at night or in diminished light conditions. Their life is precarious at best. People like to eat them—at least in the South—and they are high on the cuisine delights of fish and small mammals like the otter and mink. As for their own eating habits, nearly anything will do—dead or alive. They are not above a crayfish meal. Later in this exercise we will examine stomach contents for clues to their diet.

Procedure: External Anatomy

1. Place a preserved specimen dorsal side up on a dissection pad.

   The body plan for the crayfish is composed of a head of 5 fused segments, a thorax of 8 and an abdomen of 6 segments (**Figures 7-1 and 7-2**). Note that the body is divided into an anterior cephalothorax and posterior abdomen. The abdomen consists of several segments and is terminated by the telson, an extension of the last abdominal segment. The chitinous exoskeleton offers good protection for the crayfish, but must be shed (molting) to permit growth. The head and thorax are covered by the carapace. The line of fusion of the carapace between the head and thoracic region is called the cervical suture or groove. The rostrum is the pointed anterior portion of the carapace. It provides some protection for the eyes.

2. Note that the compound eyes are moveable on stalks. Examine one under a dissecting scope. The small, square facets on the surface are each a visual unit of the eye called an ommatidium.

3. Tail appendages:

   Crayfish appendages are conventionally numbered anterior to posterior (**Figure 7-3**), but for our purposes they are easier to keep tract of beginning at the posterior end. Carefully remove them from the left side of the animal and place them in sequence. Remember, we are counting backwards for the number of the appendage, beginning with the telson, then swimmeret 5, 4, 3 etc. Note that the appendages are considered homologous as they are similar in origin and development. In this arthropod the appendages are specialized for different functions. This is called serial homology, that is, the adaptation of a longitudinal series of originally similar structures to perform different functions. **In every case as you examine these appendages, take a moment to study them with top light under your dissecting microscope.** These are wonderful examples of matching form and function. Table 7-1 provides a summary of the appendages.

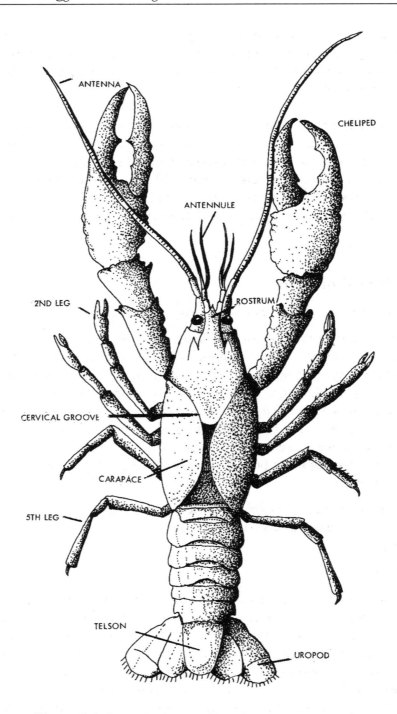

Figure 7-1. Dorsal View of Crayfish External Features

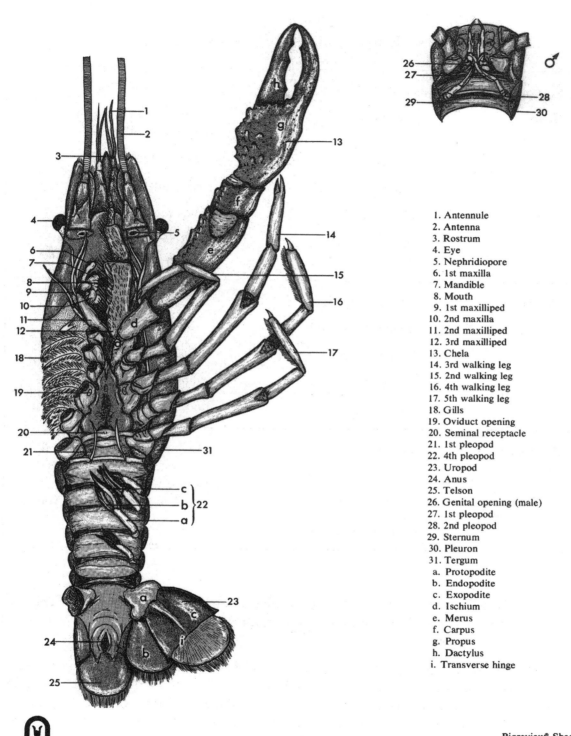

1. Antennule
2. Antenna
3. Rostrum
4. Eye
5. Nephridiopore
6. 1st maxilla
7. Mandible
8. Mouth
9. 1st maxilliped
10. 2nd maxilla
11. 2nd maxilliped
12. 3rd maxilliped
13. Chela
14. 3rd walking leg
15. 2nd walking leg
16. 4th walking leg
17. 5th walking leg
18. Gills
19. Oviduct opening
20. Seminal receptacle
21. 1st pleopod
22. 4th pleopod
23. Uropod
24. Anus
25. Telson
26. Genital opening (male)
27. 1st pleopod
28. 2nd pleopod
29. Sternum
30. Pleuron
31. Tergum
a. Protopodite
b. Endopodite
c. Exopodite
d. Ischium
e. Merus
f. Carpus
g. Propus
h. Dactylus
i. Transverse hinge

Carolina Biological Supply Company, Burlington, North Carolina 27215

Printed in U.S.A. © 1969 Carolina Biological Supply Company

Bioreview® Sheet
4452

Figure 7-2. Ventral View of Crayfish External Features

Table 7-1
Function of Crayfish Appendages

Region	Appendage	Function
Head	Antennule	Senses of touch, chemical taste and equilibrium
	Antenna	Touch and taste
	mandible	Crush and hold food
	first maxilla	Holds food at mouth
	second maxilla	Food handling and moves food from gill chamber
Thorax	first maxilliped	Touch, taste and holds food
	second maxilliped	
	third maxilliped	
	Cheliped	Holds food, defense
	Walking leg	Locomotion
Abdomen	Swimmerets	Male uses first swimmeret to transfer sperm to female. Female uses swimmerets 2–5 to carry eggs and young
	Uropod and Telson	

a. Uropods: Note the broad, biramous (2 part) uropods and central telson. This unit functions for backward swimming, and in females, to protect eggs and young attached to the swimmerets.

b. Swimmerets (Pleopods): The basic function of swimmerets is to create water currents over the gills. They also have reproductive functions. In the female, the first swimmeret may be reduced or absent. The second to the fifth carry eggs and the young. In the male, the first and second swimmerets are fused to form a trough-like channel used to transfer sperm to the female receptacle.

4. Appendages of the Cephalothorax:

a. Walking Legs: There are five pairs of walking legs located on the ventral side of the abdomen. The male genital openings of the sperm ducts are located at the base of

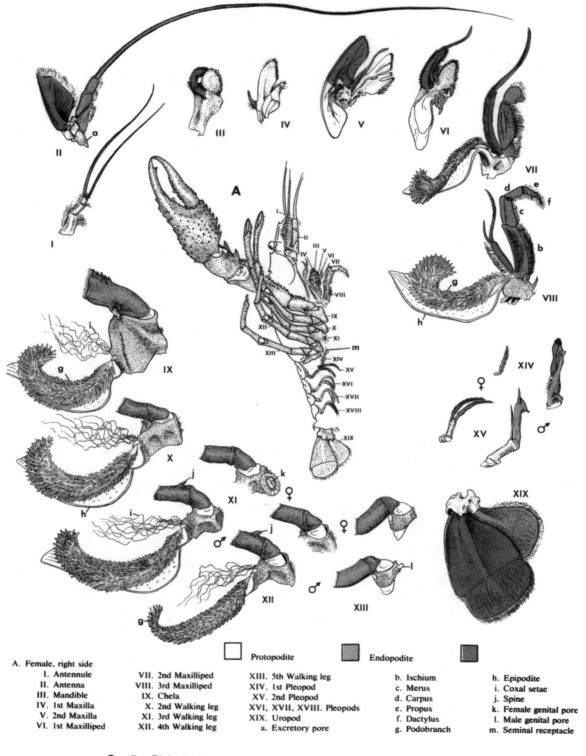

A. Female, right side
  I. Antennule
  II. Antenna
  III. Mandible
  IV. 1st Maxilla
  V. 2nd Maxilla
  VI. 1st Maxilliped

  VII. 2nd Maxilliped
  VIII. 3rd Maxilliped
  IX. Chela
  X. 2nd Walking leg
  XI. 3rd Walking leg
  XII. 4th Walking leg

  XIII. 5th Walking leg
  XIV. 1st Pleopod
  XV. 2nd Pleopod
  XVI, XVII, XVIII. Pleopods
  XIX. Uropod
    a. Excretory pore

  b. Ischium
  c. Merus
  d. Carpus
  e. Propus
  f. Dactylus
  g. Podobranch

  h. Epipodite
  i. Coxal setae
  j. Spine
  k. Female genital pore
  l. Male genital pore
  m. Seminal receptacle

Protopodite   Endopodite

Carolina Biological Supply Company, Burlington, North Carolina 27215
Printed in U.S.A. © 1979 Carolina Biological Supply Company

Bioreview® Sheet
4454

Figure 7-3. Crayfish Appendages

each of the fifth walking legs. In females, the oviduct openings are at the base of each third walking leg. The opening of the seminal receptacle in the female is midventral between the fourth and fifth pair of walking legs (see **Figure 7-4**). Be sure to locate these openings. Note that Pair 5 is not attached to a gill. Pair 1 has chelipeds with enlarged chelae used in defense and food procurement. The main functions of these appendages are walking and food handling.

b. Maxillipeds: There are 3 pairs, each attached to a gill. These appendages also function in food handling.

c. Maxillae: Crayfish have two pairs of these very small head appendages. The so-called gill bailer is located on the second maxilla. On this appendage, the terminal exopod forms a long blade that beats to move water into the gill chamber. The rather small first maxilla functions in food handling around the mandibles. The action of the maxillae may be observed on the live crayfish when it is held facing you submerged. Ask your teaching assistant for a demonstration. There is some risk here, so this observation is strictly optional!

d. Mandibles: The mandibles are on either side of the mouth. They bear teeth on the inner edges. The mandibles group and direct food into the mouth or hold it while the maxillipeds shred it. The mandibles have a firm muscle attachment. Removing them for further examination requires some care to avoid damage to the specimen.

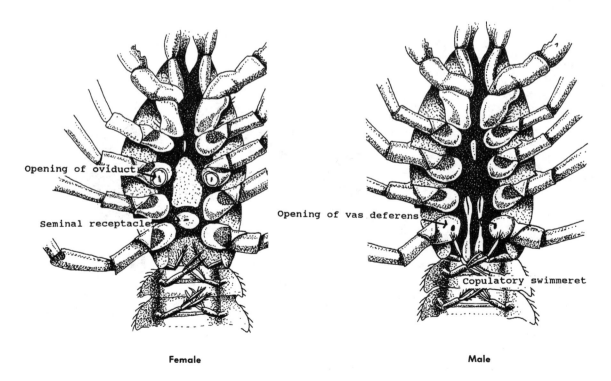

Figure 7-4. Ventral View of Walking Legs, Female and Male

e. Antennae and Antennules: These organs have both tactile and chemical sensory functions. At the base of each antennule is a statocyst important in maintaining equilibrium. This structure is located on the dorsal surface of the basal, most proximal segment. Each statocyst consists of a sac, opening to the outside. The opening is a small slit visible under the dissecting scope. It is surrounded by a prominent tuft of hairs which helps in finding this structure.

f. **With the assistance of the instructor locate the statocyst. Begin by transversely cutting away the rostrum with scissors at the level of the eyes. With forceps gently rock the antennule back and forth to sever it at the base. Examine the dorsal aspect of the antennule under the dissecting scope.** All of the above procedures are best done under low magnification. The statocyst is basically a sac containing rows of minute sensory setae arranged in a horseshoe-shaped fashion on the floor of the sac. Sand grains collected by the crayfish(!) are placed in the sac where they become attached by mucus to the setae. The setae are innervated by a branch of the antennulary nerve. The contact of the sand grain with the statocyst hairs determines the orientation of the crayfish. The statocyst is shed during molting time. At that time, animals display very little ability to orient to gravity. Gravity perception is restored as the statocyst develops and the crayfish collects new grains of sand.

Immediately after molting it is possible to place crayfish in a suspension of iron filings that become incorporated into the statocyst. The crayfish then orients to a magnetic field—until the next molt.

g. With scissors or a razor blade make a transverse cut cross-section just distal to the slit opening of the statocyst. Under 10x some of the internal cellular detail of the statocyst may be seen. This is not always successful, but worth a try.

Procedure: Internal Anatomy

1. Our first procedure is to remove portions of the carapace. Be careful with this step. The pericardial cavity, and certain main arteries lie just under the carapace and tend to adhere to it.

   Insert the scissors under the posterior edge of the carapace about 1 cm to the side of the dorsal midline. Make a longitudinal cut forward about to the level of the back cervical groove. Repeat the same procedure on the other side. Then make a transverse incision connecting the two longitudinal cuts. With forceps gently lift off the carapace beginning at the anterior end and lifting backwards. As you lift watch to make certain the underlying membranes are not pulled away with the carapace section. A probe is helpful for this procedure. Care must be taken here since the pericardial cavity is right under the section of carapace being removed.

2. **Respiratory System:**

   The exchange of oxygen and carbon dioxide is achieved by the maintenance of water over gills or branchiae in gill chambers. As shown in **Figure 7-5,** crayfish gills are enclosed in chambers beneath the sides of the carapace. These carapace regions are

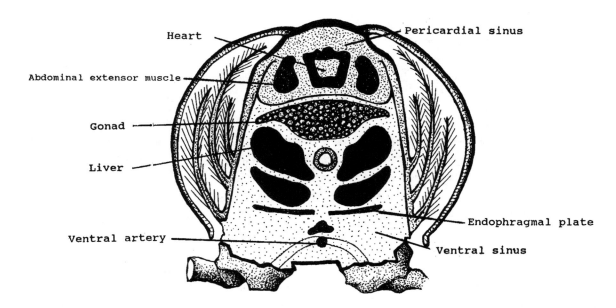

Figure 7-5. Diagrammatic transverse section through thorax of the crayfish. Dense dotted space shows oxygenated arterial blood with less dense dots indicating the region for venous blood.

called brachiostegites. In all, there are 17 gills crowded into each chamber; 6 in the outermost row and 11 attached inside. Remember that water flows over the gills by the action of the gill bailers on the second maxilla.

a. Now remove one of the gills and suspend it in water in a small petri dish. Dissecting scope study will reveal some of the details of this beautifully designed structure.

The gill more complex than it appears. Blood, after circulating through the body, collects in the ventral sinus of the thorax and abdomen (**Figure 7-5**). The blood then moves into the gills. The structure of the gill is shown in **Figure 7-6**. Within the gill, blood low in O_2 flows distally through the afferent canal (afc) and returns through the efferent canal (efc). With some attention to detail, it is possible to locate these channels in cross-sections made of the gill.

b. Place a gill on a glass slides and make thick cross-sectioned slices. Examine under the dissecting scope with bottom light. The afferent canal follows the outer side of the gill shaft, the efferent is on the inner side. The two channels are connected at the apex of the gill. The gill is designed in a manner that results in afferent blood (low in O_2) cycling through all 4 filaments of the gill before circulating to the efferent chamber and out of the gill.

3. **Circulatory System:**

Crayfish blood is a nearly colorless liquid. It contains blood corpuscles or amoebocytes. The respiratory pigment is a hemocyanin. Like other animals, the functions of the blood in crayfish are (1) to transport food and metabolic wastes and (2) to deliver O_2 from

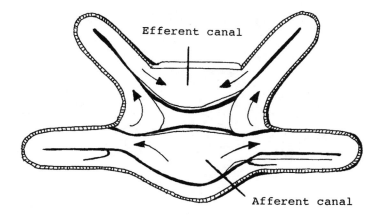

Figure 7-6. Gill Structure of the Crayfish

the gills to body tissues in exchange for CO_2. Crayfish blood will coagulate forming a clot when the animal is wounded or if, for some reason, an appendage is broken off.

The heart is a slightly muscular organ which alternately dilates to allow blood to enter through three pairs of openings called <u>ostia</u>, then contracts to push the blood out. Six ligaments, two anterior, two posterior and two running along the ventral border of each lateral surface attach the heart walls to the pericardium. These are easily torn in the dissection procedure, but some may still be in place. There are three pairs of heart openings, termed ostia. One is dorsal and the others are lateral.

a. **Locate these the ostia (or at least some of them!). Then with fine forceps remove the heart.**

b. **Under the dissecting scope it is possible to make a cross section of the heart bisecting the two lateral ostia. Place the heart on a glass slide and slice across it with a sharp razor blade. Hold your half section of the heart and through the dissecting scope look for the ostia tissue flaps on the inside wall.** These valves keep the blood

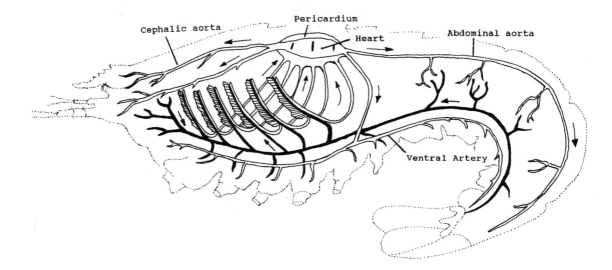

Figure 7-7. Diagrammatic View of the Circulatory System in the Crayfish

from back flowing when the heart contracts. Your dissection skills will be challenged for this procedure.

4. **Reproductive System:**

As already noted, crayfish are dioecious. Gonads are located beneath and to the anterior end of the pericardial cavity. As shown in **Figure 7-9,** the male testis consists of three lobes and two vas deferentia. These open at the bases of the fifth pair of walking legs. The paired ovary has the same location, but tends to be somewhat larger and easier to find. Oviducts lead out to the openings at the bases of the third walking legs.

a. **Locate the gonads on your specimen. The dorsal position of the pericardium should be carefully cut away exposing the gonads.**

b. Female crayfish: The appearance of the ovary will vary depending on the maturation stage. If eggs are mature they will be distinguishable as spherical orange bodies.

   Some eggs may be detached from the ovary membrane due to damage during dissection. A younger stage of ovary will have less color and look more like testis lacking, however, the prominent coiled vas deferentia. Oviducts are difficult to see. They will appear as flat, somewhat wider ducts than the vas deferens. A slide preparation of ovary unfortunately reveals little detail. Students should exchange preps to examine both sexes.

c. Male crayfish: Like the female ovary, the testis will be located midline in the thoracic cavity, directly beneath and in front of the heart. In appearance, the testis has a smooth appearance and is whitish in color. The vas deferens display prominent coiled tubes lying at the side of the testis before descending along the outside body wall to the genital opening.

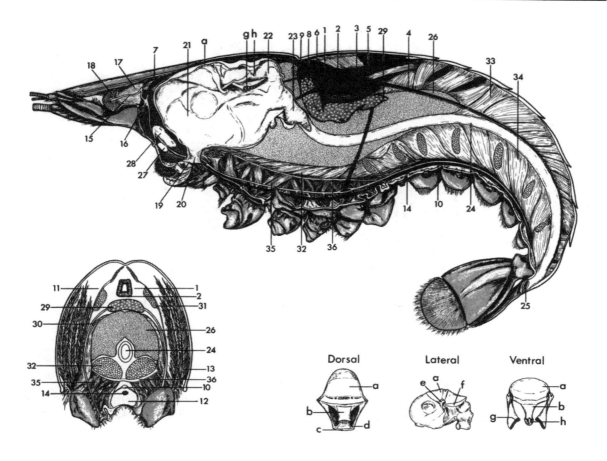

1. Pericardial sinus	19. Mouth	Gastric mill
2. Heart	20. Esophagus	a. Cardiac ossicle
3. Ostium	21. Cardiac stomach	b. Urocardiac ossicle
4. Superior abdominal artery	22. Pyloric stomach	c. Pyloric ossicle
5. Sternal artery	23. Midgut	d. Prepyloric ossicle
6. Opthalmic artery	24. Intestine	e. Zygocardiac ossicle
7. Cor frontale	25. Anus	f. Pterocardiac ossicle
8. Antennary artery	26. Liver	g. Lateral tooth
9. Hepatic artery	27. Green gland	h. Median tooth
10. Ventral nerve cord	28. Bladder	
11. Brachiocardiac sinus	29. Ovary	
12. Ventral sinus	30. Oviduct	
13. Gills	31. Extensor of thorax	
14. Ventral abdominal artery	32. Flexor of thorax	
15. Optic nerve	33. Extensor of abdomen	
16. Cerebral ganglion	34. Flexor of abdomen	
17. Eyestalk	35. Proximal claw muscle	
18. Eye	36. Endophragmal shelf	

Carolina Biological Supply Company, Burlington, North Carolina 27215

Printed in U.S.A. © 1966 Carolina Biological Supply Company

Bioreview® Sheet
4456

Figure 7-8. Crayfish Dissection

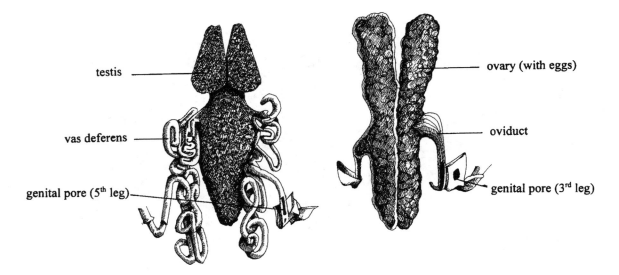

testis

vas deferens

genital pore (5th leg)

ovary (with eggs)

oviduct

genital pore (3rd leg)

Figure 7-9. The Reproductive Structures of Crayfish, Male (l.) and Female (r.)

d. Remove the testis and prepare a slide wet mount. Examine under 4X for gross details. Then remove the coverslip and macerate the tissue for examination under 10 and 40X objectives. Similar to the squid, various stages of sperm development within morula can be observed. After maturation, spermatozoa appear as flattened, spherical cells in the testis or vas deferens. However, living sperm in water uncoil, finally becoming star-shaped. They are quite unique and beautiful. Also, like squid sperm, they are packaged into spermatophores before release.

The mating behavior, both of crayfish and its cousin the lobster, is a predictable and interesting event. It is initiated by the male crayfish. In males, the first or sometimes only the second two pairs of pleopods are modified for transferring spermatophores. During the mating ritual, spermatophores are loaded in the male pleopods. Next the male turns the female over and holds her in this position with his legs and powerful tail. The female's genital pores are located on her third legs and below the pore is a seminal receptacle opening. The spermatophores are mixed with mucus and stuck to the seminal receptacle.

Fertilization is achieved later. The female lies on her back and discharges a sticky secretion with the eggs. At the same time, sperm are released from the spermatophores. The fertilized eggs become cemented to the pleopods. After these events, the female rights herself and maintains a secretive behavior while the young begin development. Juveniles hatch in several weeks depending on water temperatures. Most juveniles remain attached to mom until after the first molt. They then begin their independent life.

5. **Digestive System:** As already mentioned, crayfish are omnivorous. Their diet includes tadpoles, frogs, insect larvae and small fish. As already mentioned, most of the feeding is at night or at dusk and daybreak.

The digestive system is shown in **Figure 7-8.** Briefly, it is composed of the mouth, a short esophagus, a stomach of two parts (the pyloric and cardiac chambers), the midgut, intestine, and the anus. Like many crustacea, the crayfish stomach has special structural adaptations for grinding the food much like the function of the jaws in vertebrates.

Muscles attached to the outside stomach walls facilitate coordinated grinding movements. Inside, a spectacular arrangement of teeth and ossicles are pulled together and then apart to macerate food materials. The laterally and centrally positioned units of ossicles are termed the gastric mill. It is better seen than described.

a. Cut the remainder of the dorsal carapace away up and across the rostrum at the eye level. Expose the thoracic region of the digestive system and carefully excise the complete stomach.

b. Cut open the ventral side of the stomach and remove the contents for later microscope analysis.

c. Wash out the contents of the stomach. Pin the stomach open with dorsal side down. With good top light and low magnification some of the structural details of this extraordinary organ will be apparent to you. Compare your dissection with Figure 7-8.

The grinding of food is achieved when the gastric muscles contract stretching the stomach so that it becomes longer and narrower. This brings the lateral cusps together and swings the dorsal tooth forward to mesh and grind the food. Enzymes from the digestive gland are also released at this time. **Locate the digestive gland or liver.** A cardiopyloric valve and numerous hairs or setae serve to filter the food mash, straining it so that a mostly digestible liquid and small particles move to the pyloric chamber of the stomach while coarse residues are regurgitated out the mouth. Final straining takes place in the pyloric chamber. Note the large numbers of setae on the ventral surface. As the stomach is stretched the sides of the chamber are pressed inward and the mash is squeezed to the ventral surfaces of the stomach where rows of setae effect a final separation of food particles. Stomach ulcers are unknown for crayfish.

6. **Muscular System:** The most powerful muscles of the crayfish are those of the abdomen related to the backward swimming motion. Both the large flexor and the smaller extensor reach into the thoracic cavity as shown in **Figure 7-10**.

a. Using your scissors remove a dorsal section of the tail exoskeleton exposing both the abdominal extensor and flexor muscles.

b. Other muscles to locate are the posterior and anterior gastric muscles and the paired mandibular muscles which have been pulled away from their attachment to the carapace.

It is interesting to compare how muscles function in crayfish to how they work in vertebrates. The "skeleton" of the crayfish is external and tubular except in the ventral part of the thorax. Muscles are internal and attached to the inner surface of the skeleton. In vertebrates, of course, muscles are external to the skeletal system.

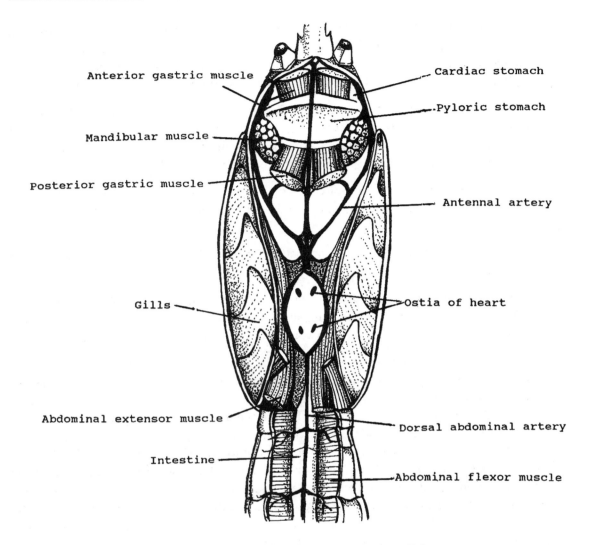

Figure 7-10. Dorsal View of the Crayfish

7. **Excretory System:** Nitrogenous compounds, mostly ammonia, are the principal waste products in crayfish. These by-products diffuse away in the gills and other regions where the cuticle is thin. The <u>antennial or green glands</u> primarily function in osmoregulation. The freshwater habitat for crayfish is slightly less saline than crayfish blood. To prevent blood dilution, the green glands eliminate excess water. The gland has distinct regions; an end sac composed of a labyrinth of walls where diffusion mostly occurs, then a convoluted tubule and finally a bladder that drains through a short duct to the outside. The exit opening was seen on the basal coxa of each of the antennae.

   a. **The green glands can be viewed after removal of the stomach. However, do not attempt to remove them as doing so will disturb the nervous system.** Dissection of green glands for internal structure is not feasible in preserved specimens.

8. **Nervous System:** We complete our study of the crayfish by looking at the nervous system. It is this system more than any other that reflects the capabilities of the animal and the crayfish is no exception. Survival represents a behavior fine-tuned to its environment with activities coordinated by the brain and its neurons.

   The crayfish nervous system is much like that of the annelid, but clearly with a more complex and larger combination of dorsal cerebral ganglia (**Figure 7-11**). There are four pairs of ganglia in this structure. The most anterior two pair (joined) innervate the eyes, the next pair has nerves going to the first antennae and the last pair innervate the second antennae. Connectives then pass ventrally around the esophagus to the ventral nerve cord. The ventral process is a double nerve cord with ganglia in every segment, although in many segments this is not apparent because they are closely fused.

   Crayfish sometimes survive an encounter by rapid retreat generated by the large flexor tail muscles. Enervation for this action comes from giant nerve fibers within the ventral nerve cord. The cell body for the nerve is located in the brain or in a ventral ganglion and then runs posteriorly. Rapid muscle contractions require elevated nerve stimulation. Remember the giant axons of the squid necessary for its rapid predatory and escape movements generated by contraction of the mantle.

   a. **For this dissection remove the tail exoskeleton by making two longitudinal cuts with scissors and lifting off the exoskeleton. Also, remove any remains of the intestine and digestive glands, and gently wash out the thoracic cavity.**

   b. **Pin the crayfish with dorsal side up. Begin at the tail and with a pair of scissors and cut through the muscles on both sides avoiding damage to the nerves below. Remove the muscles with a pair of forceps.**

      You will see that in the abdomen the two central nerve cords are fused. As you work your way into the cephalothorax the central nerve cords will disappear from view. Actually, they are covered by the endophragmal skeleton, a series of inward extensions of the exoskeleton for the attachment of the thoracic muscles. The endophragmal skeleton, in turn, is covered by longitudinal muscles.

   c. **Remove the longitudinal muscles by cutting through their posterior ends and lifting them up with forceps. If any lateral muscles should be in your way remove them in the same fashion.** The endophragmal skeleton should be exposed at this point. **Carefully remove it bit by bit with fine scissors and forceps, again working your way from the posterior to the anterior end, cutting as closely as possible to the thoracic wall.** Slowly, the central nerve cord will be exposed until it links up with the sub-esophageal ganglia connectives. From these ganglia connectives encircle the pharynx to fuse with the dorsal cerebral ganglia. Nerve processes leading from the brain to the eyes and antennae can be seen.

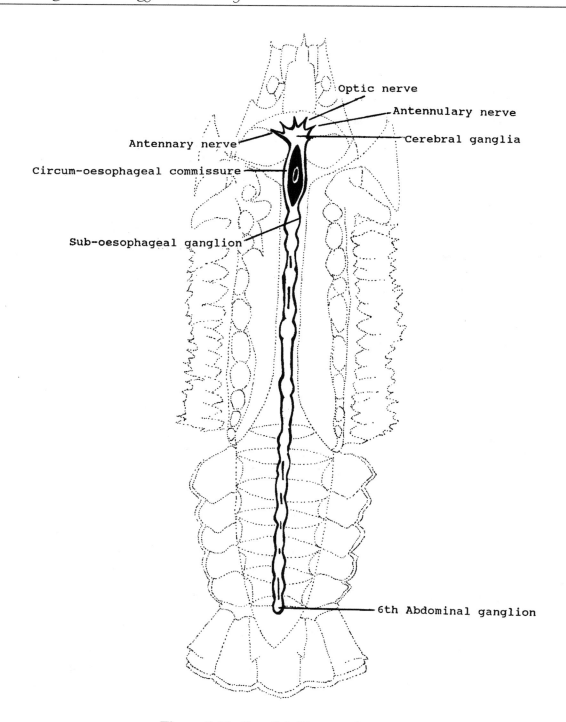

Figure 7-11. Crayfish Nervous System

Grasshopper Dissection:

Procedures: External Anatomy

(Refer to Figures 7-12 to 7-14 at the end of this section for help in identifying structures.)

1. Get a lubber grasshopper from the jar. The grasshoppers were fixed in formalin, and there are gloves available to handle them. Wash the grasshopper off and put in a pan.

2. Using the figures provided examine the external anatomy of the grasshopper. Note the **head, thorax,** and **abdomen.** The head is specialized for sensing the environment and feeding. It also acts as a relaying center to send sensory information to the rest of the body. The head is composed of five fused segments, called **sclerites,** with paired appendages. Observe the compound eyes which are very similar to those of the crayfish except they are not held on stalks. Do you think this affects the grasshopper's vision?

3. Note the 3 **ocelli** or simple eyes that detect the presence or absence of light. Two antennae sense vibrations. The grasshopper mouthparts are adapted for biting and chewing. They are covered by a bilobed flap, **labrum.** Lift this upper lip and expose the **mandibles** that are used to chew leaves. Note that they are heavily sclerotized (hardened) and move in a lateral plane. Each mandible is closed by a powerful muscle. Remove the labrum, mandibles and the **paired maxillae.** The lower lip or **labium** consists of fused second maxillae. The mouthparts have been evolutionarily modified in different insects according to their method of feeding. Butterflies, for example, have mouthparts consisting of a long proboscis that acts as a tube through which nectar is sucked up. When not in use the proboscis is tightly coiled. What is the function of each mouthpart? Which parts are palpi and what is their function?

4. The three thoracic sements (**prothorax, mesothorax,** and **metathorax**) each bear a pair of legs. Look at the attachment points of the legs on the thorax. The three thoracic segments are in the form of a box composed of four schlerites: on the dorsal side is the **notum,** on the sides are **pleura,** and on the ventral side is the **sternum.** Locate the thoracic spiracles, which are the openings to the tracheal system, which allows direct gas exchange with the tissue.

5. Next, look at the wings. The front wings are thickened into a **tegmen** and function to protect the membranous hind wings. Note how the wings are attached so that they can pivot and fold over the back. This was a major evolutionary innovation of the neopterous (new wing) insects. Remove the wings so you can continue dissection.

6. Remove a leg and identify the segments: **coxa, trochanter, femur, tibia, tarsus,** and pretarsus with **claws.**

7. Look at the abdomen; the dorsal surface is called the **tergum** and the ventral surface the **sternum.** Locate the abdominal spiracles. Count the number of segments in the abdomen. The abdomen differs between the sexes. The males have blunt abdomens and females have pointed **ovipositors** used for egg laying. What sex is your grasshopper?

Compare external characteristics of the crayfish (crustacean) and grasshopper (insect). Using the table below:

Characteristic	Crustacea	Insecta
Body divisions:	cephalothorax and abdomen	head, thorax, abdomen
Number of paired Appendages:		
Antennae		
Mouthparts		
Legs		
Gas Exchange:		
Principal habitat:		

Procedures: Internal Anatomy

1. Remove the head and using a scalpel perform a sagittal section to expose the brain (supraesophogeal ganglion).

2. Place the grasshopper right side up and carefully cut through the exoskeleton with your scissors. Place some water in your dissecting tray, which will help float the organs, particularly the trachea which will look like a mass of branched silvery tubes. If you are careful you will see the heart, which is a dorsal tube, although often it gets removed with the exoskeleton. Notice the many muscles that are attached to the exoskeleton, try to follow some and discern their purpose. The space between the body wall and the internal organs is the **hemocoel** that has replaced the **coelom.**

3. Locate the digestive system. If your grasshopper is female everything may be covered with a large yellow mass of ovaries, follow them back to the ovipositor. Then push the ovaries aside and look for the gut. The **esophagus** leads from the mouth to a sac-lik structure, the **crop** that functions to store food. Under the esophagus you may be able to see very small grayish grape-like clusters of the **salivary glands.** The esophagus leads to the **midgut** that is covered with **gastric caeca,** that function in storage. These are grainy triangle-shaped objects on top of the gut. Follow the gut back to the **hindgut** where the very slender **malpighian tubules,** are attached. These act as a kind of kidney for removing nitrogenous wastes in insects.

4. Remove the digestive system and look for the ventral nerve cord and ganglia this will appear as two white lines along the inside of the sternum that are periodically connected with larger white masses. The numerous ganglia control movement of single segments or a small body region. Also, look again for the extensively branching **tracheal system.** The tracheae are networks of passages connecting the surface of the grasshopper to all portions of its body. Oxygen diffuses through these passages directly to the cells without the intervention of the circulatory system. This kind of respiratory system is effective only for small animals.

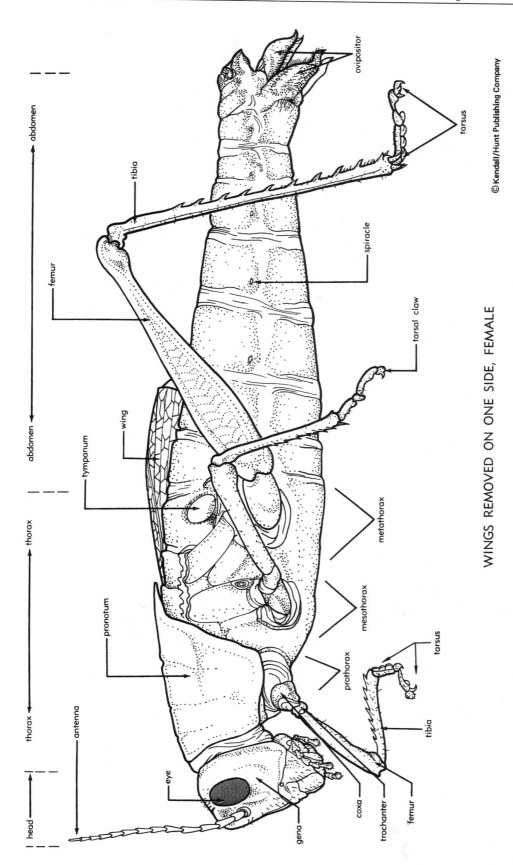

WINGS REMOVED ON ONE SIDE, FEMALE

Figure 7-12. Grasshopper, External Anatomy Lateral View

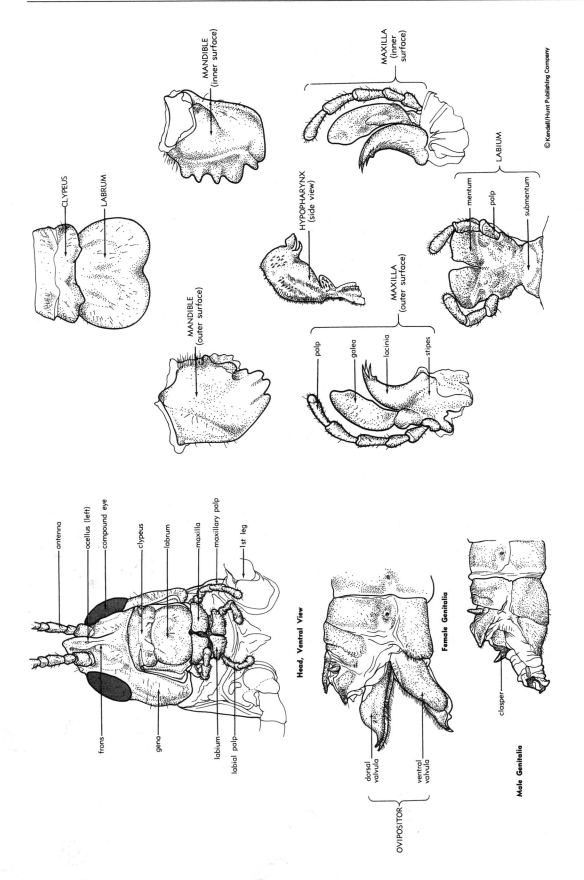

Figure 7-13. Grasshopper, External Anatomy, Mouthparts and External Sex Characteristics

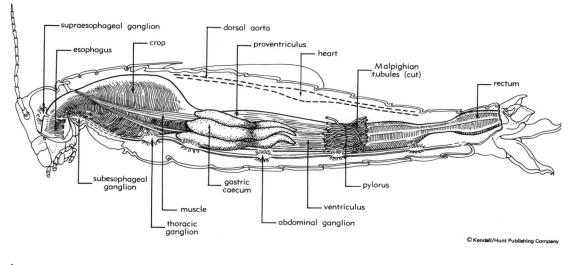

© Kendall/Hunt Publishing Company

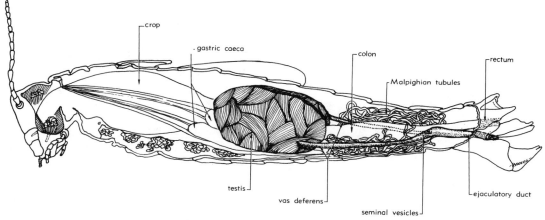

© Kendall/Hunt Publishing Company

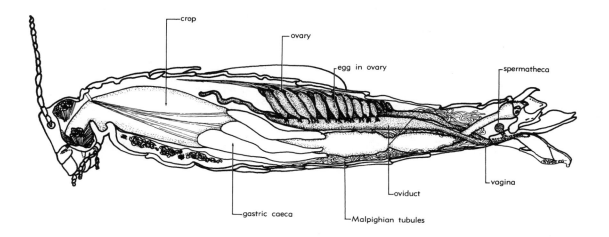

Figure 7-14. Grasshopper, Digestive Tract and Reproductive Systems

Crayfish Checklist

External Anatomy

Antennule
Antenna
Rostrum
Nephridiopore
Maxilla
Mandible
Maxilliped
Chela
Walking Leg
Gills
Oviduct Opening
Seminal Receptacle
Uropod
Anus
Telson
Opening of Green Gland
Opening of Statocyst
Ommatidium

Internal Anatomy

Pericardial Sinus
Heart
Ostium
Ventral Nerve Cord
Gills
Cerebral Ganglion
Eyestalk
Cardiac Stomach
Pyloric Stomach
Intestine
Anus
Liver
Green Gland
Vas Deferens
Testis
Spermatozoa
Egg
Ovary
Extensor of Thorax
Flexor of Thorax
Extensor of Abdomen
Flexor of Abdomen
Gastric Mill
Anterior and Posterior
Gastric Muscle

Grasshopper Checklist

External Structures

Head
Thorax
Abdomen
Ocelli
Eye
Labrum
Mandibles
Maxillae
Maxillary palp
Labium
Labial palp
Prothorax
Mesothorax
Metathorax
Coxa
Trochanter
Femur
Tibia

Tarsus
Claws
Ovipositor

Internal Structures:

Hemocoel
Esophagus
Crop
Gizzard
Gastric caeca
Midgut
Hindgut
Malpigian tubules
Ovary
Oviduct
Testis
Rectum
Brain
Ventral nerve cord

References

Bliss, D.E. 1982–1985. The biology of crustacea. New York. Academic Press, Inc.

Cameron, J.N. 1985. Molting in the blue crab. Sci. Am.

Cronin, T.W., N.J. Marshall, and M.F. Land. 1994. The unique visual system of the mantis shrimp. Amer. Sci.

Emerson, M.J., and F.R. Schram. 1990. The origin of crustacean biramous appendages and the evolution of the Arthropoda. Science.

Hadley, N.F. 1986. The Arthropod Cuticle. Sci. Am.

Hickman, C.P., L.S. Roberts, and A. Larson. 1996. Integrated Principles of Zoology. Wm. C. Brown.

Pearse, V., Pearse, J., Buchsbaum, M., Buchsbaum, R. 1987. Living Invertebrates. Blackwell Scientific Publications and The Boxwood Press.

Schmitt, W.L. 1965. Crustaceans. Ann Arbor, The University of Michigan Press.

Schram, F.R. 1986. Crustacea. New York, Oxford University Press.

Phylum: Chordata
Dissection of the Fetal Pig

Exercise
8

Several species of animals are used in laboratory courses to provide students with a better understanding of the anatomy and function of organ systems in chordates. We have chosen the fetal pig. Specimens are readily available from the meat industry, have good size, and much of the structure of this mammal is similar to our adult anatomy or to fetal structures like the umbilical cord and circulation. The gestation period is about 113 days. At birth, pigs are 12–14″ long, with 7 to 8 offspring in a litter.

General Instructions

For good success, a dissection must be carried out systematically and with care. Our purpose is to expose tissues and organs in their normal positions with as little structural disruption as possible.

Dissection considerations: The best strategy to approach any dissection is by being well-prepared. Read the exercise over and study all figures and illustrations. As each part of the dissection is done, use structural "landmarks" that are easy to identify to pinpoint the location of other more obscure structures. Use the dissecting lamps provided to give better illumination of small structures and cavities and use a probe to identify structures before cutting with a scalpel. A needle or probe can also be used effectively in any procedure that involves tissue removal or membrane peeling. Sharp scalpel blades are very helpful. New blades are recommended for each dissection lab.

A pig will be provided to each pair of students. Protect your hands from preservative by wearing gloves. Rinse the preservative from the pig and dry it with paper towels. Further directions will be given by your TA. At the conclusion of the lab, clean the bench area thoroughly.

External Features

1. **Note** the umbilical cord functioning to connect the fetus with the placenta.

2. The anus is just ventral to the tail.

3. **Sex determination.** Female pigs are easier to identify due to distinctive external <u>urogenital papilla</u> located ventral to the anus. In males, the preputial orifice of the penis is located just caudal to the attachment of the umbilical cord. Two paired swollen areas that are spongy to touch are the immature <u>scrotal sacs</u> located below and ventral to the anus. Be sure to **compare** your fetal pig with one of the opposite gender.

Internal Features

Digestive System

The digestive system includes the oral cavity, salivary glands, esophagus, stomach and intestines. Because of time constraints we will concentrate on the anatomy of the pig starting from the neck and continuing posteriorly

Oral Cavity

Force the jaw open slightly. Take your scalpel and cut through the cheek and masseter muscle (ckeek muscle) on both sides. The jaw bone or mandible should be pried open until it dislocates. It may be necessary to slide your thumb far into the throat to generate leverage and to use force. The <u>hard palate</u> makes up the roof of the mouth nearest the nostrils. (See **Figure 8-1.**) It is a rigid structure with tiny bumps that make it possible to breath and chew food at the same time. The soft palate located posterially allows mammals to suckle by making an airtight seal. The nasopharyngeal "opening" is where the nasal cavity connects with the throat (<u>pharynx</u>) and

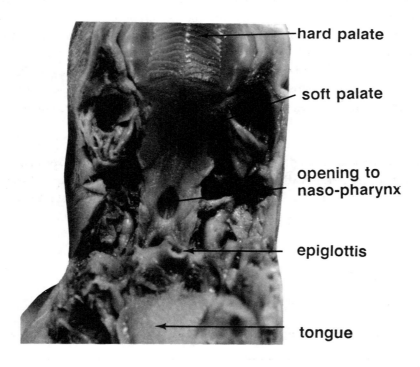

Figure 8-1. Oral Cavity

is located on the ventral side of the tongue. Further back, a projecting flap of tissue, the <u>epiglottis</u> may be found. The epiglottis closes off the <u>glottis</u> (the opening into the larynx) preventing food and water from entering into the respiratory tract. Off to the side of the mouth it may be possible to locate the submandibular glands. Finally, introductory dentistry may be practiced by removing one or more of the molar teeth. Cut into the jaw with your scalpel, remove the flap of skin, and pry out the molar (optional procedure).

Abdominal Cavity

A pair of 2–3 foot pieces of twine should be cut and one end of each piece securely tied to the fore and hind feet on one side. The pig should then be placed in the dissecting pan, both pieces of twine brought underneath the pan and tied to the opposite fore and hind feet. It may be easiest to simply wrap it around the feet several times so that it may be easily tightened later on.

Note: The incisions in the head and thoracic region must not be too deep, or veins and arteries may be accidentally cut. Refer to **Figure 8-2** for the placement of incisions. The goal of all the cuts, unless otherwise specified, is to remove the skin without cutting into the muscles.

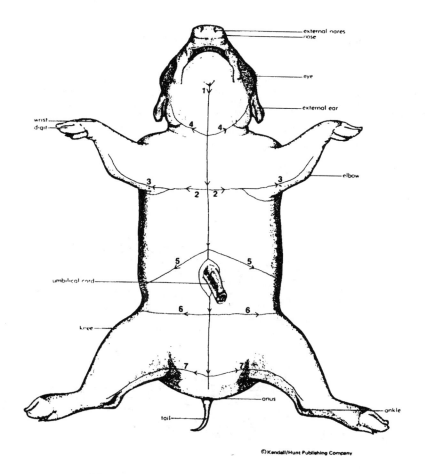

Figure 8-2. Ventral View for Incisions

Peritoneum

The abdominal cavity is separated from the thoracic cavity by the <u>diaphragm</u>. In order to open the abdominal cavity, **palpate the ribs determining where the thoracic and abdominal cavities are located.** Refer to **Figure 8-2** in order to **make the opening incisions in the abdominal region.** The incisions must not be too deep or the underlying organs will be cut. The remaining portion of the abdominal cavity is surrounded by a thin semi-opaque membrane known as the <u>peritoneum</u>. If the skin was carefully removed the peritoneum may still be present. If you are looking directly at the abdominal organs, then the peritoneum is attached to the skin. The skin flaps should be peeled off and cut away.

The part of the peritoneum lining the abdominal cavity is called the <u>parietal peritoneum</u>. Another membrane called the <u>visceral peritoneum</u> covers the visceral organs such as the intestines, stomach, liver etc. The potential space between the peritoneum and visceral peritoneum is the <u>peritoneal cavity</u>, which is part of the embryonic coelom. **The peritoneum should now be removed** to fully expose the abdominal organs. The abdominal cavity is often filled with fluid and spilled dye making it difficult to observe structures. Therefore, **the abdominal cavity should be briefly rinsed** with running water.

Visceral Organs

The brown, four lobed <u>liver</u> is the most prominent feature of the abdominal cavity (**Figure 8-3a, b**).

However, it may be obscured by the umbilical cord. Therefore, **cut the umbilical cord vein and move the umbilical to one side. Identify** the large left lateral lobe, the left central lobe, the right central lobe, and the small obscured right lateral lobe. Gently lift the right central lobe in order to identify the gall bladder which has a green appearance. The exact appearance of the liver will vary from animal to animal since the liver changes its shape to fill up any cavity.

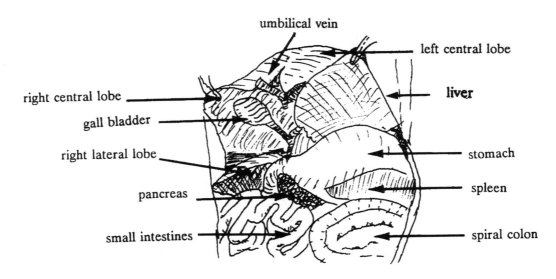

Figure 8-3a. Ventral View of Visceral Organs

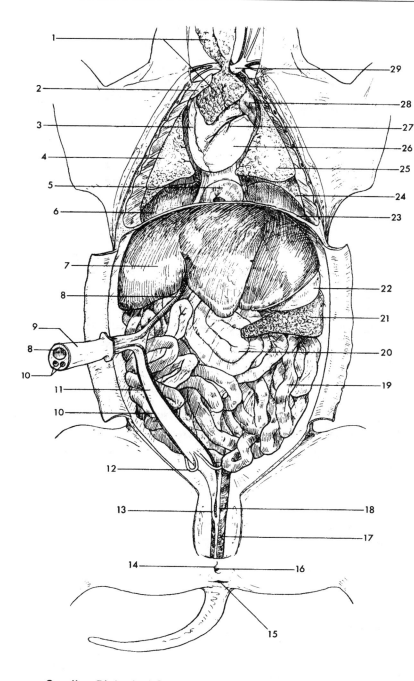

1. Thymus
2. Lung, right apical lobe
3. Pericardium, partly removed
4. Lung, right cardiac lobe
5. Lung, right intermediate lobe
6. Lung, right diaphragmatic lobe
7. Liver
8. Umbilical vein
9. Umbilical cord
10. Umbilical artery
11. Urinary bladder
12. Ureter
13. Urethra
14. Genital papilla
15. Anus
16. External urogenital orifice
17. Rectum
18. Vagina
19. Small intestine
20. Large intestine
21. Spleen
22. Stomach
23. Diaphragm
24. Lung, left diaphragmatic lobe
25. Lung, left cardiac lobe
26. Left ventricle
27. Coronary artery and vein
28. Left atrium
29. Aortic arch

Carolina Biological Supply Company, Burlington, North Carolina 27215
Printed in U.S.A.© 1969 Carolina Biological Supply Company

Bioreview® Sheet

Figure 8-3b. Fetal Pig Anatomy

The gall bladder should now be examined in greater detail. (See **Figure 8-4**.)

Locate the fibers that run from the gall bladder to the small intestines just below the stomach. This is the gastrohepatic ligament which should be removed. **Carefully scrape it away using your probe** to reveal the common bile duct. The common bile duct empties into the small intestines or duodenum just below the stomach. Closer examination will reveal that the common bile duct is made up of the cystic duct which comes directly from the gall bladder and the hepatic duct from the liver.

Digestive System

The stomach is located under the left lateral lobe of the liver. **The stomach should be removed** by locating and cutting the esophagus. The esophagus passes through the diaphragm and is often covered by the liver. Also associated with the stomach is a thin, long, dark red organ called the spleen. The spleen is involved with storage, destruction of red blood cells and the formation of white blood cells. **The stomach is attached to the spleen by the gastrosplenic ligament which should be cut. The small intestine near the stomach or duodenum should also cut. It may be necessary to cut away other connective tissue in order to remove the stomach. Using a pair of scissors cut the stomach open.** A green or clear fluid may flow out. This fluid is a mixture of mucus and merconium.

Identify the cardiac valve which allows food to enter into the stomach from the esophagus. If it functions improperly, so called "heartburn" results from the release of stomach acid into the esophagus. The pyloric valve is a circular muscle which controls the passage of food out of the stomach into the duodenum.

Looking into the space left by the stomach's removal, **identify** the pancreas. The soft, cream-colored, lobular pancreas is located between the stomach and the duodenum. The pancreas produces insulin, glucagon, and pancreative juice containing amylase, trypsin, peptidase, and lipase. Unfortunately, locating the pancreatic duct which connects to the duodenum is difficult.

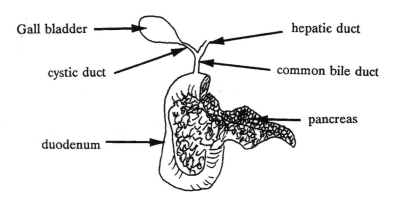

Figure 8-4. Gall Bladder

The small intestine is composed of the duodenum, jejunum and ileum. These segments all look identical and are held together by a clear delicate membrane called the mesentery. The visceral membrane, previously discussed, contains the mesenteries. The duodenum comprises the first 1–2 cm of the small intestine. The jejunum and ileum are indistinguishable except by microscope examination. On the inside, both sections have a velvety appearance due to villi, which increase the absorptive surface area.

The large intestine can be identified by both the larger diameter and **green color**. The large intestine begins with the spiral colon that is coiled tightly together. Vitamins, minerals, and water in the alimentary canal are reabsorbed in the large intestine. Feces and a larger number of bacteria are stored in the rectum, or the final part of the large intestine. **In order to observe the rectum, remove the small and large intestine by cutting the mesenteries.** The rectum will usually still be attached to the lower abdominal cavity. The rectum passes through the muscles found in the pelvis and terminates at the anus.

Excretory System

The kidneys serve as a blood-filtering organ in order to remove metabolic waste and maintain osmotic balance. The urine flows out of the kidney through a small, tan tube called the ureter. **Carefully follow the ureter until it enters the urinary bladder.** In the fetal pig, the urinary bladder is attached to the umbilical cord and has a long "upright vacuum bag" shape. In the female pig, the ureter leaves the bladder and joins the vagina where it exits the body at the urogenital sinus. In the male, the ureter is barely visible as it enters into the penis.

Remove one kidney by cutting the ureter and blood vessels. **Make a longitudinal cut** to display the internal structure. **Identify** the outermost striated cortex. Bowman's capsule, the proximal tubules and the distal tubules are all located in the cortex. The loop of Henle extends downward into the medulla. In the longitudinal section the medulla has a smooth appearance. The urine enters into the central pelvis from the collecting tubules. From the pelvis the urine then exits out of the kidney through the ureter.

Female Reproductive System

Posterior to each kidney is a small, oval, bean-like ovary (see **Figures 8-5a and 8-5b**).

The mesentery suspends both ovaries away from the dorsal wall of the abdominal cavity. Coming from each ovary is a small tube known as the oviduct or fallopian tube. The body (or horn) of the uterus is formed where the two oviducts come together. Fertilization typically occurs in the fallopian tubes and development occurs in the uterus. In order to expose all of the uterus and vagina, it will be necessary to cut through the pelvic bone with your scalpel. Keep the hindlegs apart by tightening the string on them. With external examination alone, it is difficult to differentiate the uterus from the vagina. However, the two are separated by a construction called the cervix. Therefore, it is possible to insert a probe into the anterior uterus and move it posteriorly towards the vagina to find the cervix. The cervix will stop the probe's movement. The uterus will be anterior and the vagina posterior to this point.

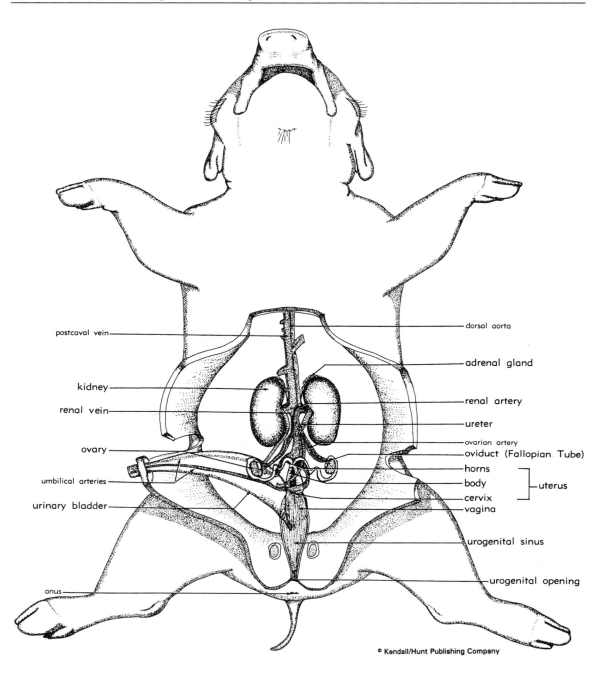

© Kendall/Hunt Publishing Company

Figure 8-5a. Female Urogenital System

Male Reproductive System

The penis is closely associated with the umbilical cord. **Locate** the urogenital orifice of the penis just behind the umbilical cord (see **Figures 8-6a and 8-6b**). The tubular (posterior) penis can be felt by pressing the tissue associated with the umbilical cord and applying a slight rolling force. **Use your probe to free the penis from the connective tissue.** In the adult, the penis only extends from the surface during copulation.

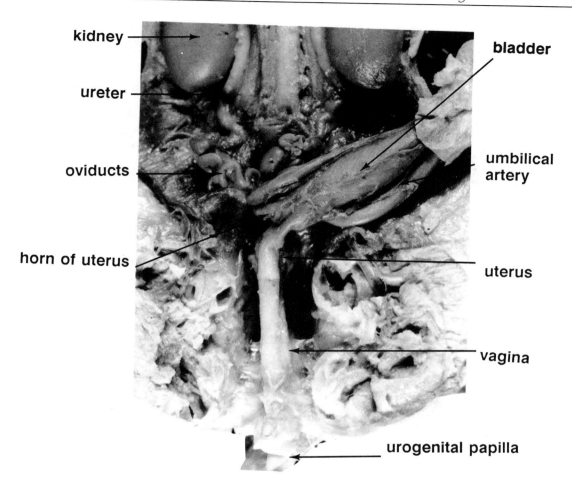

kidney

ureter

oviducts

horn of uterus

bladder

umbilical artery

uterus

vagina

urogenital papilla

Figure 8-5b. Female Urogenital System

Locate the scrotal sacs ventral to the anus. **Carefully cut the scrotal sacs open** to expose the testes, epididymis, and inguinal pouch. The testes are ventral to the epididymis, which is an extremely coiled tubular structure. Sperm pass from the testes into the epididymis where they are stored and undergo further maturation. The inguinal pouch is a clear sac that is prominent in the fetal pig.

Cardiovascular System and a Few Interesting Glands

Begin by **identifying** the pectoralis muscles. These muscles connect the sternum to the forelegs. They serve as an important landmark for locating a network of blood vessels and nerves directly underneath. **Start by cutting into the axillia area ("armpits") just posterior to the pectorals.** It should be possible to locate the network of vessels and nerves going into the forelegs after this cut is made and with the use of the probe. In order to better expose the network, **remove the superficial and the deep pectoral muscles a layer at a time.**

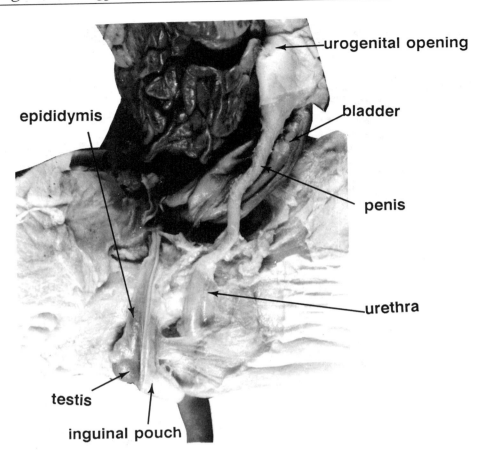

Figure 8-6a. Male Urogenital System

The scissors should be used to cut through the ribs just lateral of the ventral midline. The midline cut is avoided so that the sternum does not have to be cut in half. The scissor cuts through the ribs should be shallow and parallel to the skin to avoid a deep cut into the heart. Continue cutting until all of the ribs have been severed. In order to avoid damaging the network of vessels entering the foreleg, cut parallel to the ribs just posterior to the vessels. After making this cut, peel the ribs back, using some force, and then cut the flaps away.

Some of the ribs and pectoralis muscle should still be present. **The remainder of the pectoralis muscle should be removed. It will be necessary to whittle through the sternum with a sharp scalpel.** After part of the sternum has been removed **identify** the sternomastoid muscle. This muscle starts on the mastoid process (just behind ear) and attaches to the sternum. **The sternomastoid muscle must be removed in order to completely remove the sternum.** Upon removing the muscle it will be possible to visualize the thymus gland. At this point, continue to remove the sternum. Eventually all that will remain of the sternum is a connection between the second pair of thoracic ribs. **These should be carefully pried away to completely expose the thoracic cavity.**

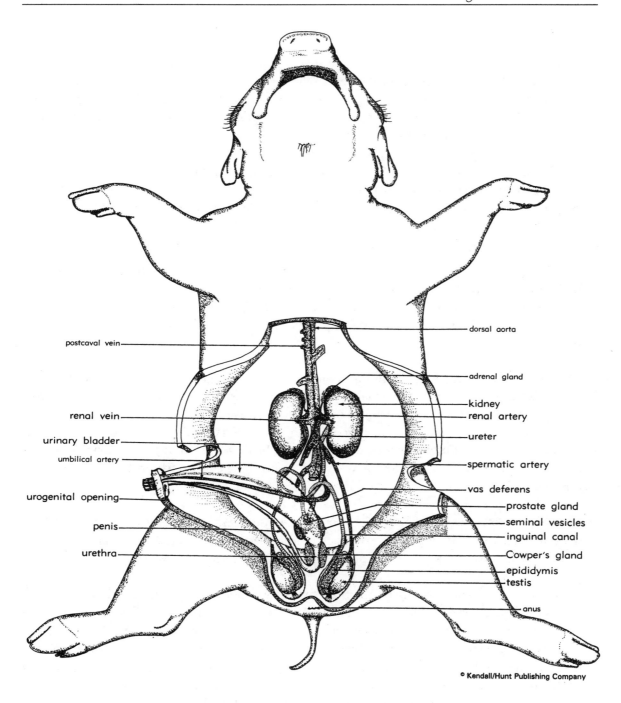

Figure 8-6b. Male Urogenital System

Thymus and Thyroid

The <u>thymus gland</u> is a prominent feature that begins in the fetal pig at the larynx and continues posterially to cover the top of the heart. The thymus is similar in color and consistency to the pancreas. Although the thymus is prominent in the fetal pig after puberty it atrophies and may all but disappear in adults. The thymus plays an important role for the production of lymphocytes which help produce antibodies.

Covering the thymus gland in the midline is the <u>sternothyroid muscle</u>. If the sternothyroid muscle is **gently** opened a dark brown, hard oval gland is seen in the midline. This is the <u>thyroid gland</u> produces the hormone thyroxin that affects metabolism, growth, and differentiation.

Anterior Veins

In order to expose the anterior veins, **it will be necessary to remove the thymus and thyroid gland. It is important to carefully remove the thymus** since both the artery and veins are located underneath. The thymus should also be removed from the heart. It is easiest to remove the pericardial membrane that covers the heart and then remove the thymus that is located outside the membrane.

Veins have been dyed blue by an injection of blue latex. Veins are defined as vessels bringing blood towards the heart. Unfortunately the latex occasionally leaks, especially in the throat region that are slit to allow the latex injection. Arteries are dyed red. The arteries are smaller and occasionally also have some blue coloration due to the dye mixing. In order to differentiate the two, a small hole with a probe may be drilled into the blood vessel. If any red is found, then the structure is an artery. For the study of veins and arteries consult **Figures 8-8 and 8-9** and the supplemental figure at the end of the exercise.

In order to better expose the anterior veins, **it will be necessary to remove connective tissue with your probe.** In order to expose the brachial artery, it may be necessary to start at the foreleg and clean connective tissue with a probe while working back to the heart.

Once sufficient connective tissue has been removed it should be easy to **identify** the <u>anterior vena cava</u>, the large vessel that enters into the right atrium of the heart. This vein drains the head, neck and arms. **Trace** this vein forward and note that it is formed by the union of the two left and right, <u>brachiocephalic veins</u>. **Trace** the right brachiocephalic vein forward. It collects blood from the <u>internal jugular vein</u>, the <u>external jugular vein</u> and the <u>subclavian vein</u> that comes from the forelegs. It may be possible to **trace** the subclavian vein through the chest wall where it branches into the <u>subscapular</u> and <u>axillary veins (also called the brachial vein)</u>.

Anterior Arteries

The <u>dorsal aorta</u> arises from the left ventricle of the heart. The <u>aortic arch</u> is the portion of the aorta that curves to the pig's left. As it descends to the posterior, it is called the <u>descending aorta</u>. To see the descending aorta, it may be necessary to cut the upper left lobe of the lung and push the heart to the right. The first vessel exiting from the aortic arch on the right side is the <u>brachiocephalic trunk</u> or artery. This divides <u>right subclavian artery</u> and the right and left

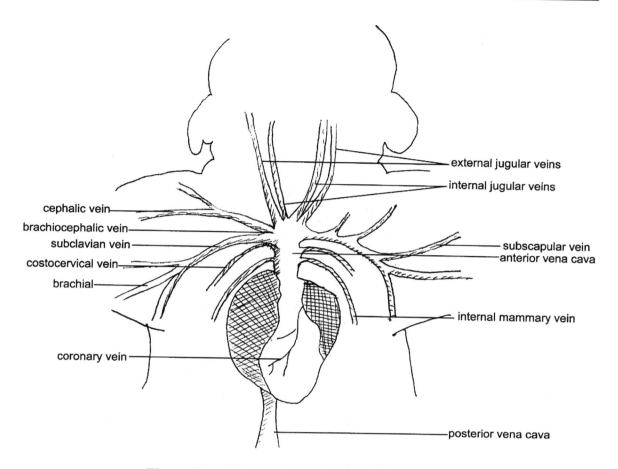

Figure 8-7. Fetal Pig. Anterior Veins, Ventral View

common carotid arteries. The subclavian artery passes through the pigs axilla (armpit) region to become the axillary artery, then the brachial artery on the upper arm. The common carotid supplies the brain and head with oxygen and is just lateral of the trachea. Do not confuse the carotids with the thin, white vagus nerve.

The second artery to leave the aortic arch is the left subclavian. It also becomes the left axillary and brachial artery.

If the heart is pushed to its right it should be possible to see the pulmonary artery which comes from the right ventricle. The pulmonary artery splits into the left and right pulmonary arteries that then go to each lung. In the fetal circulation, the lungs do not function and the blood coming from the right ventricle is already oxygenated. The ductus arteriosus allows the blood to flow directly from the pulmonary artery into the descending aorta. In the fetal pig, the ductus arteriosus is more prominent than the left and right branches of the pulmonary artery. At birth, when the lungs begin to function, pressure changes cause the ductus arteriosus to close almost immediately.

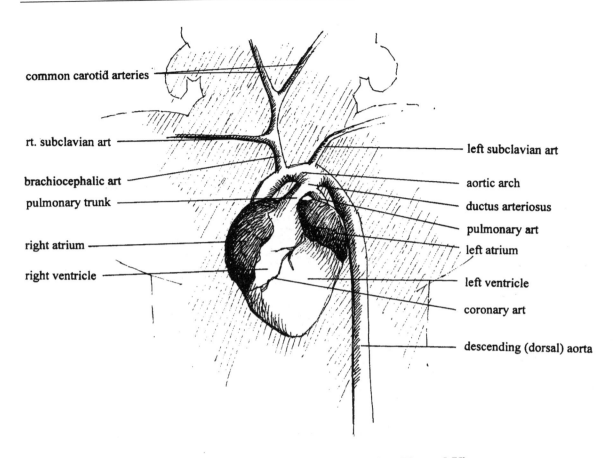

common carotid arteries

rt. subclavian art

brachiocephalic art

pulmonary trunk

right atrium

right ventricle

left subclavian art

aortic arch

ductus arteriosus

pulmonary art

left atrium

left ventricle

coronary art

descending (dorsal) aorta

Figure 8-8. Fetal Pig. Anterior Arteries, Ventral View

The heart should be carefully removed for later examination. Any remaining pericardial membrane should also be removed. The posterior vena cava, and all connecting arteries should be cut. It may also be necessary to cut connective tissue.

Abdominal Cavity

In order to expose the veins and arteries found in the abdominal cavity, fold the diaphragm down. Cut the umbilical vein and move it out of the way. Push the liver to its right in order to expose the stomach. Remove the stomach by cutting the esophagus, duodenum, and bile duct. In addition, the entire small intestines and colon should be removed.

Posterior Veins

The large, blue posterior vena cava located in the midline and descending through the abdominal cavity is the largest blood vessel seen. The kidneys serve as a distinguishing landmark and the renal veins are easily located. As the posterior vena cava continues posterially, it is found underneath the descending aorta. The vena cava then splits into the left and right iliac vein. It may require removing connective tissue in order to locate the iliac veins. Eventually, the iliac vein enters into the hindleg paralleling the course of the arteries.

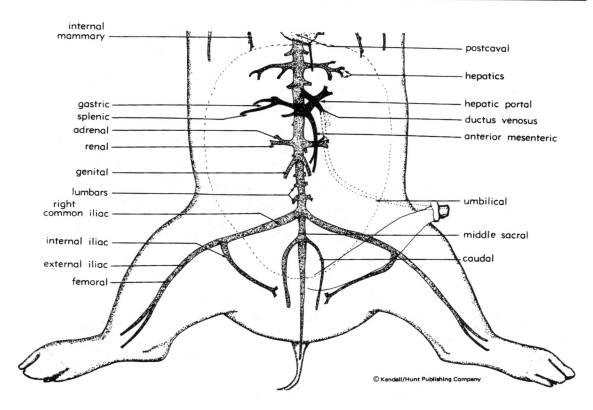

Figure 8-9. Fetal Pig. Posterior Veins

Posterior Arteries

The descending aorta is often covered by connective tissue making it difficult to locate some of the posterior arteries if care is not taken. The first artery leaving the descending aorta after passing through the diaphragm is the coeliac artery usually hidden by considerable connective tissue. By identifying the artery supplying the spleen or liver it is possible to work backwards to the coeliac artery. It follows that the two branches of the coeliac are called the gastrohepatic (to liver) and the gastrosplenic (to spleen).

Toward the posterior along the descending aorta, the next artery located is the mesenteric. Usually in the process of removing the small and large intestines most of the mesenteric artery is removed. The artery normally enters the mesentery and then branches out. However, upon careful examination, it is possible to locate this artery.

The renal artery is large, posterior to the mesenteric, and is usually hidden by the renal vein. Therefore cut the renal vein and move the connective tissue with your probe.

After passing by the kidneys the descending aorta is called the abdominal aorta. It then divides into the iliac arteries. However, in the fetal pig the umbilical arteries are more prominent. The iliac arteries branch off slightly before and are more lateral than the umbilical arteries. The iliac arteries further branch into the deep femoral and the femoral. The deep femoral is more medial while the femoral can be traced with a probe down the hind leg.

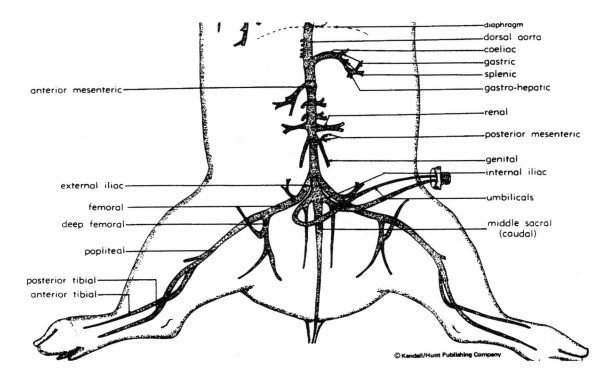

diaphragm
dorsal aorta
coeliac
gastric
splenic
gastro-hepatic
anterior mesenteric
renal
posterior mesenteric
genital
internal iliac
external iliac
umbilicals
femoral
deep femoral
middle sacral
(caudal)
popliteal
posterior tibial
anterior tibial

© Kendall/Hunt Publishing Company

Figure 8-10. Fetal Pig. Posterior Arteries

Heart

Blood enters the heart through the anterior and posterior vena cava which empty into the right auricle. The auricle will be dark brown. The holes left by the vena cava are best observed from the dorsal side. The right ventricle is directly below the right atrium. Exiting from the right ventricle is the pulmonary artery which then splits into left and right branches. The pulmonary artery is more ventral than the aorta which can easily be confused. Two holes are then observed in the left atrium where the pulmonary veins return. Below the atrium is the left ventricle which also includes the apex of the heart. The aorta then exits from the left ventricle. Depending upon your dissection, the brachiocephalic artery may still be attached.

Histology (Selected Organs)

You will also get to look at some prepared slides of cells and tissues that make up the organs of the digestive tract.

Stomach

The stomach is a dilated part of a temporary (usually about two hours) treatment center for ingested food. Gastric juices and a mechanical churning action reduce food to a liquid state called chyme. With the interesting exception of alcohol, there is little absorption of nutrients

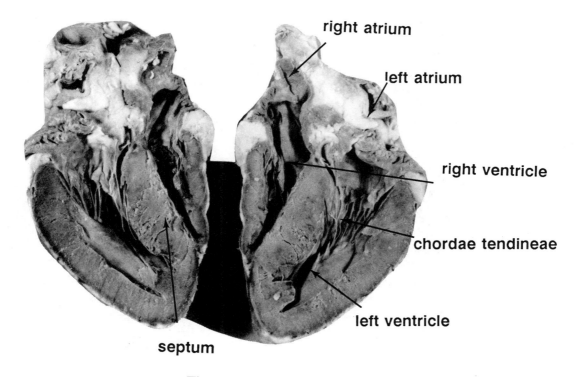

right atrium

left atrium

right ventricle

chordae tendineae

left ventricle

septum

Figure 8-11. Cow Heart Section

from the stomach. After chyme is generated, the <u>pyloric sphincter</u> relaxes, releasing chyme into the duodenum.

The histology of the stomach appropriately reflects its function. The mucosa walls are arranged into longitudinal folds called <u>rugae</u>. This allows the stomach to greatly expand as it is distended by food. At the base of the mucosa there is a well-defined musculature. The mucosa lining consists of tubular glands that synthesize and secrete the gastric juices. Two components are <u>hydrochloric acid</u> (pH 0.9 to 1.5) and the enzyme <u>pepsin</u> that hydrolyses proteins. The cells of the mucosa are protected from these secretions by a covering of <u>mucus</u>. Note this layer in the slide preparation. There are 3 basic cell types located in the tubular mucosa with specific functions.

1. <u>Mucus secreting cells</u> found mostly at the tips of the tubular mucosa. Typically the nucleus is at the base of the cell.

2. <u>Parietal cells</u>. These secrete HCl. They are large, round prominent cells and tend to be more numerous in the middle portion of the gland.

3. <u>Chief cells</u>, pepsin secreting cells which tend to be grouped at the base of the gastric glands. Strands of muscle extend along the base of the glands and between them. The muscle layers contract to expel gastric juices into the lumen of the stomach.

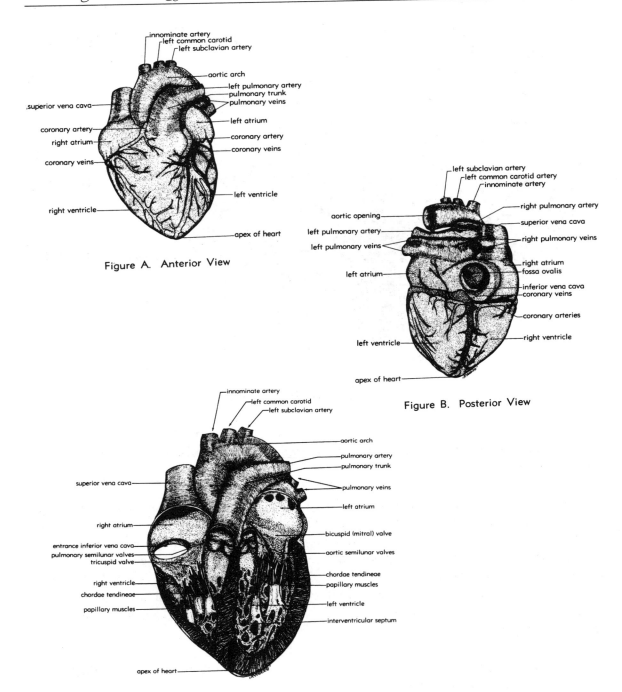

Figure A. Anterior View

Figure B. Posterior View

Figure C. Sagittal View

Human Heart. Supplemental Figure

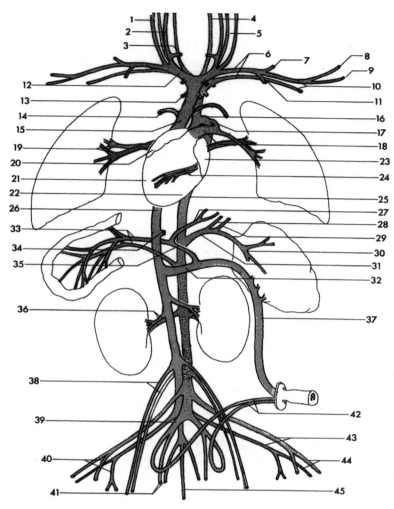

1. External jugular v.
2. Internal jugular v.
3. Thyroid v.
4. Common carotid a.
5. Thyrocervical a.
6. Subclavian a. & v.
7. Subscapular v.
8. Radial v.
9. Ulnar v.
10. Brachial a. & v.
11. Lateral thoracic v.
12. Innominate v.
13. Internal mammary v.
14. Costocervical v.
15. Precaval v.

16. Aortic arch
17. Ductus arteriosus
18. Pulmonary trunk
19. Pulmonary a. & v.
20. Right auricle
21. Right ventricle
22. Left ventricle
23. Left auricle
24. Coronary a. & v.
25. Dorsal aorta
26. Postcaval v.
27. Gastrosplenic v.
28. Splenic a.
29. Gastric a.
30. Hepatic a.

31. Coeliac a.
32. Ductus venosus
33. Superior mesenteric a. & v.
34. Hepatic v.
35. Hepatic portal v.
36. Renal a. & v.
37. Umbilical v.
38. Genital a. & v.
39. Hypogastric a.
40. Internal iliac a. & v.
41. Sacral a. & v.
42. Umbilical a.
43. Common iliac a. & v.
44. External iliac a. & v.
45. Caudal v.

 Carolina Biological Supply Company, Burlington, North Carolina 27215
Printed in U.S.A. © 1969 Carolina Biological Supply Company

Fetal Pig Anatomy 8-12: Blood Vascular System

Small Intestine

Structurally adapted for major functions of secretion and absorption, the small intestine is characterized by prominent villi that project into the lumen of the intestine. Each villus consists of simple <u>columnar epithelium</u> and scattered <u>goblet cells</u>. The interior contains both blind capillaries and lymphatics. Consult your text for details.

The small intestine secretes large quantities of mucus to protect the villi from gastric juices. The mucus is produced in cells of <u>Brunner's glands</u> which can be located deep in the mucosa. The cells typically surround a central lumen. They show a prominent basal nucleus with lightly stained cytoplasm filled with mucigen.

Kidney

The functioning units of the kidney begin in <u>Bowman's capsules</u> located in the <u>cortex</u> of the kidney. Locate them on the slide preparations. Consult your text for details.

Ovary

Main ovary functions are to produce gametes (ova) and secrete sex hormones. At the time of birth the ovaries contain thousands of primordial egg cells. Usually less than 400 ever mature and are released from the ovary. Each oocyte is enclosed by a single layer of epithelium. They originate as <u>primary follicles</u> in the outermost portion of the ovary cortex. At sexual maturity, primary follicles enlarge dramatically and push outward to the ovary surface. These are termed <u>Graafian follicles</u>. During ovulation, the outer ovary wall ruptures and the oocyte is expelled into the abdominal cavity where it is usually drawn into one of the ciliated oviducts (Fallopian tubes). Examine ovary slides for follicle development. Consult your text for details.

Bio 204 Checklist

Lab 8—Kingdom <u>ANIMALIA</u>
Phylum Chordata

<u>External Anatomy</u>:
Male: scrotal sacs
Female: papilla

<u>Oral Cavity</u>:
hard palate
soft palate
epiglottis
opening to nasopharynx
teeth

<u>Digestive System</u>:
thoracic cavity
abdominal cavity
esophagus
stomach
liver
spleen
pancreas
gall bladder
small intestine
large intestine
colon
rectum
coelom
mesenteries
cystic duct
hepatic duct
bile duct
pyloric sphincter
cardiac sphincter

<u>Urogenital System</u>:
Urinary—
    kidney
        cortex
        medulla
        pelvis
    ureter
    urinary bladder

Female Reproductive—
    ovary
    oviduct/Fallopian
    uterus
    vagina

Male Reproductive—
    penis
    inguinal pouch
    testis
    epididymis

Circulatory System:

Arteries:
aorta/aortic arch
brachiocephalic
rt/left subclavian
rt/left common carotids
ductus arteriosus
pulmonary trunk
pulmonary artery
descending aorta
renal
iliac
femoral
deep femoral
umbilical

Veins:
internal/external jugular
subscapular
anterior/posterior vena cava
brachiocephalic/ innominate
pulmonary
subclavian
axillary or brachial
renal
iliac
femoral
deep femoral
umbilical

Other:
Thymus
thyroid
sternothryroid muscle

Dissection of the Sheep Brain

Introduction

In the previous exercises, the relative simplicity of the hydra nerve net and the cerebral ganglia of the earthworm have been seen. This exercise will focus on the mammalian brain, by far the most complex organ known both in structure and function. The sheep brain will be studied. This organ is of sufficient size to permit detailed study and is readily available as a by-product of meat processing plants. The structure of the human brain is summarized in **Figure 9-1**.

Procedure

Working with a partner, obtain a sheep brain and gently rinse in water. Gloves should be worn to avoid exposure to the preservative. Examine the gross external structure to correlate your view with Figure 9-1. Figures 9-2, 9-3 and 9-4 will aid in the dissection.

Gross Anatomy Forebrain

Telencephalon

The telencephalon or cerebrum is the most prominent part of the mammalian brain. It comprises about seven-eighths of the total weight of the brain. It consists of the neocortex (cerebrum), paliocortex, and archiocortex. The neocortex is highly convoluted in order to increase the surface area of the gray matter. The paleocortex and archiocortex are more primitive parts of the cortex. The paliocortex is represented by the pyriform lobe and contains the olfactory regions. The archiocortex is completely hidden by the neocortex and will be revealed by later dissection.

The outer surface of the telencephalon or cerebrum consists of gray matter with prominent upfolds termed gyri. The deep downfolds are termed fissures. Shallow downfolds are called sulci. These folds greatly increase the surface area available for the gray matter. A longitudinal fissure divides the cerebrum into right and left halves called hemispheres. As shown in **Figure 9-2**, the neocortex is divided into the frontal, parietal, temporal and occipital regions. The central sulcus separates the frontal lobe from the parietal lobe. The lateral cerebral sulcus separates the frontal lobe from the temporal lobe. The parietocipital sulcus divides the parietal lobe from the

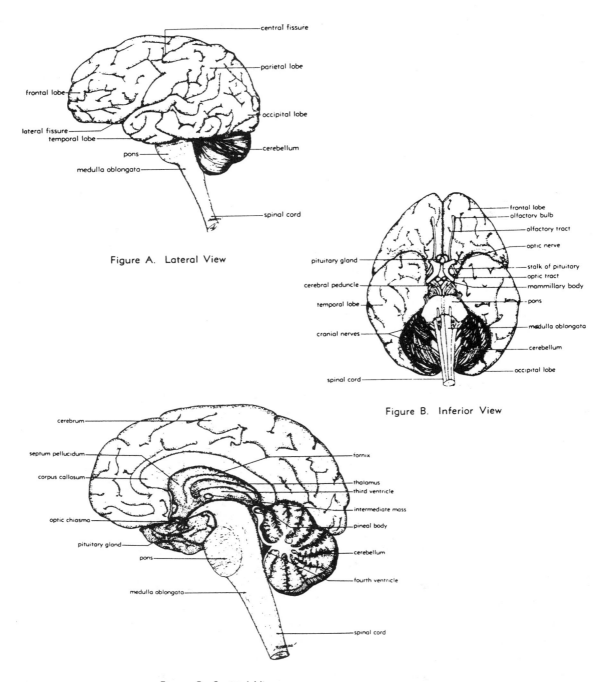

Figure A. Lateral View

Figure B. Inferior View

Figure C. Sagittal View

Figure 9-1. Human Brain

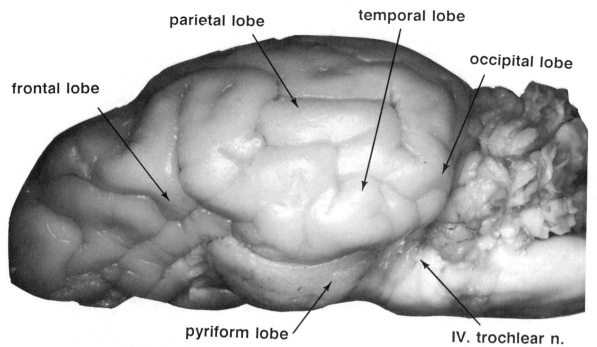

Figure 9-2. Sheep Brain. Lateral View of Neocortex

occipital lobe. The <u>transverse fissure</u> separates the cerebrum from the cerebellum. Some of the sulci are not easily distinguished in the sheep.

Identify these regions in the sheep brain.

Diencephalon

The <u>diencephalon</u> consists of the thalamus, hypothalamus and epithalmus (pineal). Turn the brain so that the ventral side is facing you. Using available illustrations, identify the prominent optic nerves. The point where the two optic tracts cross is called the <u>optic chiasm</u>. This serves as an important landmark in our exploration of brain structure. The hypothalamus is directly above the optic chiasm and posterior to it. On your specimen you may find a tiny stalk or a large cluster covered with a white/gray substance. If the large cluster is present, then the <u>pituitary gland</u>, which is also part of the diencephalon, remained attached to the brain. If only the tiny stalk remains or a small hole, then the pituitary was removed. The hypothalamus rests on top or dorsal to the pituitary stalk and pituitary. In order to observe the <u>thalamus</u>, turn the brain back over so that the dorsal side is facing you. Gently lift the posterior aspect of the neocortex up and away from the midline. Then, gently press down on the brain stem. Identify the distinctive <u>cerebellum</u>, and the thalamus.

Midbrain

Mesencephalon

The midbrain consists of the <u>superior</u> and <u>inferior colliculi</u> and the cerebral <u>peduncle</u>. Once again, position the brain so that the dorsal side is facing you and gently push down on the brain

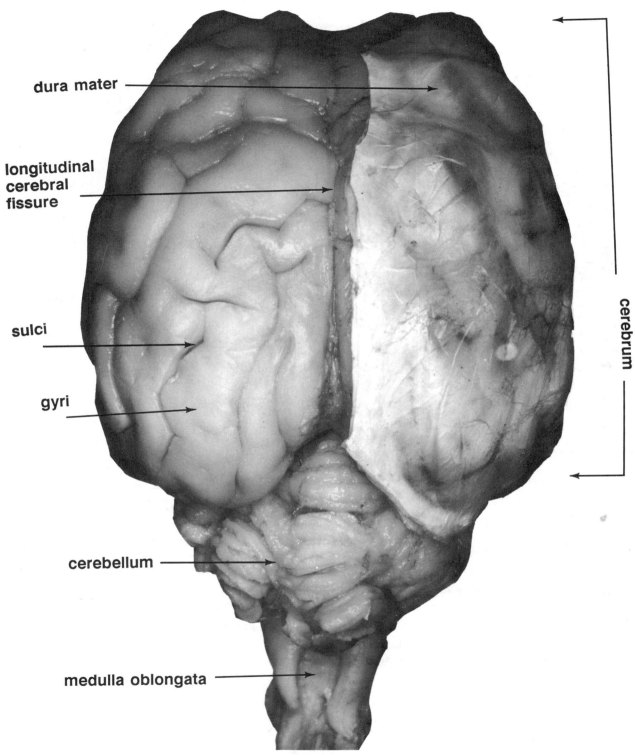

dura mater

longitudinal
cerebral
fissure

sulci

gyri

cerebrum

cerebellum

medulla oblongata

Figure 9-3. Sheep Brain. Dorsal View

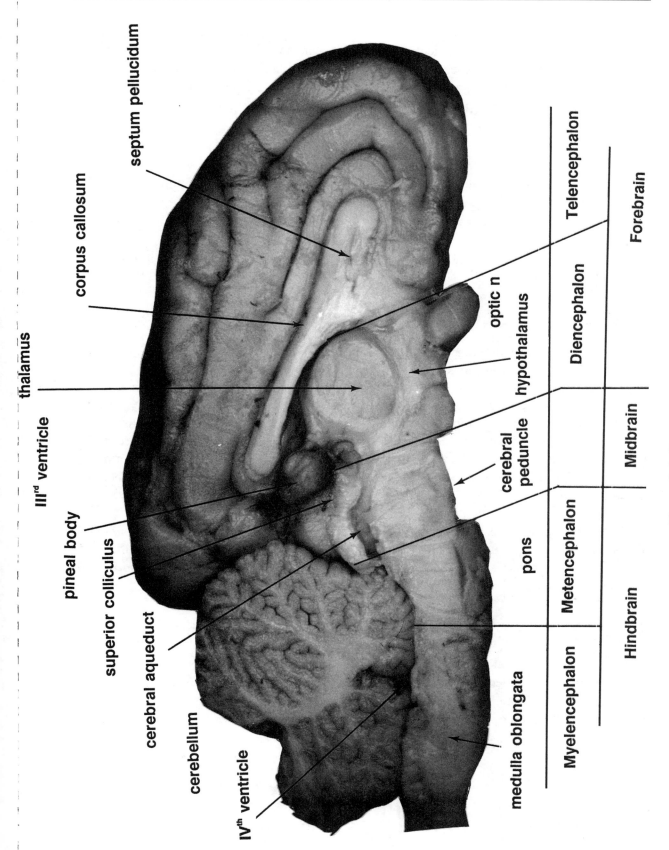

septum pellucidum

corpus callosum

thalamus

IIIrd ventricle

pineal body

superior colliculus

cerebral aqueduct

cerebellum

IVth ventricle

optic n

hypothalamus

cerebral peduncle

pons

medulla oblongata

Telencephalon

Diencephalon

Midbrain

Metencephalon

Myelencephalon

Forebrain

Hindbrain

Figure 9-4. Sheep Brain. Saggital Section

stem. Identify the two pairs of bumps between the cerebellum and the thalamus. The most anterior bump is called the superior colliculi because it appears on top of the lower bump or inferior colliculi. Next, turn the brain over so that the ventral side is facing you. The <u>cerebral peduncle</u> makes up the floor of the midbrain. The peduncle is posterior to the optic chiasm and the hypothalamus. It is anterior to the <u>pons</u> which is near the cerebellum.

Hindbrain

The anterior portion of the hindbrain consists of the <u>pons</u> and <u>cerebellum</u>. The pons is observed on the ventral side of the brain while the cauliflower-like cerebellum is observed on the dorsal side.

The medulla oblongata is in the posterior portion of the hindbrain. The medulla is best located at the ventral side of the sheep brain. The medulla is the structure posterior to the pons.

Meninges

The meninges are membranes covering the brain consisting of the <u>dura mater</u>, <u>arachnoid</u>, and <u>pia mater</u>. The dura mater is a thin but tough connective tissue that provides a barrier to injury. It is the outermost of the three layers and in the sheep brain appears as a white membrane. The dura mater is not always visible on brains preserved in ethyl glycol. The arachnoid lies underneath the dura mater. The arachnoid will not be visible with gross examination. Between the arachnoid and the pia mater is the <u>subarachnoid space</u>. It is in the subarachnoid space that the cerebral spinal fluid (CSF), and blood vessels are found. The fluid plays a major role in absorbing shock. The pia mater is a thin membrane that closely adheres to the neocortex surface.

With the tip of a probe, gently break the pia mater on the top of the cerebrum. Gently pull off a segment of the pia mater. In most cases any surface blood vessels will also be pulled off, indicating the arachnoid is also being removed. Next, place the probe in a sulcus (grove) and examine for any membrane which would also be the pia mater.

Clinical Note: Meningitis

Meningitis is an infection or inflammation of meninges, usually the pia mater and the subarachnoid space. The infection can spread rapidly throughout the CNS since the subarachnoid space is continuous. Fever, headache, and a stiff neck are common complaints. Diagnosis is made by removing CSF and looking for an increased number of white blood cells.

Cranial Nerves

The cranial nerves control our sense of sight, hearing, balance, taste, and smell (**Figure 9-5**). They also control the sensation of touch and motor control in the face. With some of the specimens, it may not be possible to identify all of the cranial nerves. If the pia mater is still attached to the ventral side, it should be left intact since removal can disturb the smaller cranial nerves. Identify as many of the cranial nerves as possible. Share your specimen with other partners if most of the cranial nerves can be found.

Olfactory Nerve (I)

This nerve will not be visible since it begins in the nose, goes through the skull and enters the olfactory bulb. However, the olfactory bulb can and should be identified.

The olfactory nerve is a pure sensory nerve which is tested by having a subject close their eyes and smell various objects.

Optic Nerve (II)

The optic nerve is the easiest of all the cranial nerves to identify due to its large size. The nerve is made up of axons from the retinal ganglion cells in the eye. The nerves terminate in the thalamus and rostral colliculus. The point where the optic nerves cross is called the optic chiasm.

The functioning of the optic nerve is tested by checking visual acuity. Severe damage to the optic nerve results in blindness.

Oculomotor Nerve (III)

The oculomotor nerve is primarily a motor nerve to the ocular muscles and terminates in the cerebral peduncle. It is fairly distinct and innervates the muscles that cause the iris to open (dilate) and close.

Its function can be tested by shining a light into the pupil. If the nerve and the brain is functioning properly both pupils should constrict. In cases of severe head trauma where the brain swells, the oculomotor nerve becomes compressed. Under these conditions, the eye remains fixed and dilated in response to light.

Trochlera (IV)

This nerve also innervates the eye and controls one of the muscles that cause the eyes to move. The nerve arises from the dorsal surface of the midbrain and can be found emerging just behind the inferior colliculus. See **Figure 9-2** for location.

Trigeminal (V)

The trigeminal nerve carries both motor and sensory information to and from the face muscles, skin, and teeth. In cases of severe chronic facial pain, it can be surgically cut. The trigeminal nerve is thick, usually cut short, and found laterally between the pons and cerebellum.

The motor aspect of the trigeminal is assessed by asking the subject to clench their teeth and feeling their masseter (jaw) muscle. The sensory aspect is tested by making sure the subject can feel light touch on the forehead, cheekbone, and jaw.

Abducens (VI)

This motor nerve is found on the ventral side of the brain originating just posterior to the pons. If the pia has not been removed, it usually is prominent being near the midline. This nerve along with the III and IV controls eye muscles. These nerves are tested by asking the subject to look up and down and sideways.

Facial (VII) and Vestibulo-Cochlear (VIII)

The VII and VIII cranial nerves are small and difficult to locate. First locate the VI cranial nerve then look laterally near the cerebellum at much the same level as the V cranial nerve. Usually only the two small stumps of the nerves remain. The facial nerve is slightly more anterior and medial. It controls head and face muscles. It is tested by having the patient frown, shut their eyes tightly, show their teeth, smile and puff out their cheeks. The VIII nerve contains axons from the auditory organs of the inner ear and balances information from the vestibular organs. The VIII nerve is tested by checking the subject's hearing.

Glossopharyngeal (IX)

Vagus (X)

Spinal Accessory (XI)

The ninth, tenth, and eleventh cranial nerves form a cluster located on the medulla. They are located laterally near the cerebellum. They are often difficult to locate or easily confused with the twelfth cranial nerve. If successful in locating the 7th and 8th cranial nerves, the 9th–11th cluster is located 1–2 mm posterially.

The glossopharyngeal nerve innervates pharyngeal muscles, and taste buds. The vagus nerve is the major parasympathetic nerve and inhibits the heart, constricts bronchi, and stimulates activity in the stomach, pancreas, and gall bladder. The spinal accessory nerve innervates muscles in the neck, shoulder, pharynx, and larynx.

Hypoglossal Nerve (XII)

The hypoglossal nerve is the most posterior of the cranial nerves and arises from the medulla. It is also more medial than the IX, X, XI cluster. The hypoglossal nerve innervates the muscles of the tongue.

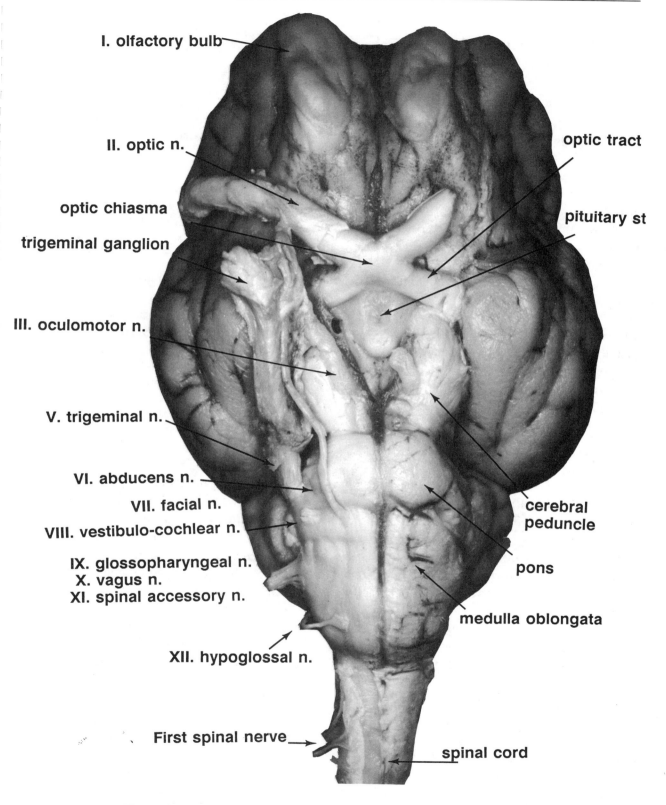

I. olfactory bulb

II. optic n.

optic chiasma

trigeminal ganglion

III. oculomotor n.

V. trigeminal n.

VI. abducens n.

VII. facial n.

VIII. vestibulo-cochlear n.

IX. glossopharyngeal n.
X. vagus n.
XI. spinal accessory n.

XII. hypoglossal n.

First spinal nerve

optic tract

pituitary st

cerebral peduncle

pons

medulla oblongata

spinal cord

Figure 9-5. Sheep Brain. Ventral View Showing Locations of Cranial Nerves

Brain Dissections

Archiocortex

In order to expose the archiocortex, more specifically the hippocampus, the brain must be dissected. The cerebellum should be removed by making two cuts with a scalpel through the cerebellar peduncles. These peduncles are located posterior to the inferior colliculus and lateral to the IV ventricle. The cuts are best made by cutting lateral to medial.

Once the cerebellum is removed, gently use a probe to expose the corpus callosum. This will require gently prying the two hemispheres open. The corpus callosum is a white fibrous collection of myelinated and non-myelinated axons that connect the neocortex of the two hemispheres together.

At approximately the level of the optic chiasm cut just one of the cerebral hemispheres down to the level of the corpus callosum. The tissue should be removed by gently scraping it away with the blunt side of your scalpel handle. It will also be necessary to remove the pyriform cortex. In order to expose the hippocampus, the corpus callosum must also be gently removed. Below the corpus callosum is the lateral ventricle (see below) with a highly vascular structure called the choroid plexus. The lateral ventricle should be pulled open in order to expose the hippocampus. This structure is named for its curving seahorse shape. Its believed to be involved with memory and emotions.

Mid-sagittal Section

The brain has a series of interconnected hollow spaces. They are filled with cerebrospinal fluid and are called the ventricles. To exam the ventricles, make a mid-sagittal cut separating the two hemispheres. If the pituitary is still attached, cut it off at the pituitary stalk (infundibulum). A long, thin knife is best for this cut; otherwise use a sharp scalpel. Place the brain ventral surface down. With a smooth continuous stroke cut between the longitudinal fissure, through the corpus callosum and down through the brain stem. Avoid jerky, sawing motions. Two completely identical mid-sagittal sections should lay before you (seldom happens).

Since a mid-sagittal section does not always occur exactly in the midline, the two sections may be slightly different. It is common that in one section a large hole exists underneath the corpus callosum while on the other a thin walled structure exists. Occasionally the thin walled structure will be seen on both sections underneath the corpus callosum. It is called the septum pellucidum and separates the lateral first and second ventricles.

The roof of the lateral ventricles is the corpus callosum. The lateral ventricles open into the third ventricle which is a space separating the right and left halves of the hypothalamus and thalamus. The third ventricle is connected to the fourth ventricle via a narrow passageway through the midbrain called the cerebral aqueduct. The fourth ventricle is located between the cerebellum, pons and medulla oblongata. Examine the ventricles for the choroid plexus. It will appear as a purple grayish tissue with many blood vessels. It functions to produce the cerebrospinal fluid.

Locate the <u>thalamus</u> and as many other structures as you can.

The <u>pineal body</u> is a small oval body, gray in color, located posterior and dorsal to the thalamus. It usually is cut in half, but sometimes it is only found on one of the halves. The pineal gland plays a role in gonadal function until puberty when it begins to calcify. In sparrows, it functions in the detection of light and control of circadian rhythms.

The <u>hypothalamus</u> is differentiated from the thalamus by a small groove. As its name implies, it rests underneath the thalamus. In the sheep, it is posterior to the optic chiasm.

The hypothalamus serves as the control center for homeostatic processes. It controls thermoregulation, hunger, thirst, mating behavior, the "fight-or-flight" response, and pleasure. It also controls the hormones released in the pituitary gland.

The hypothalamus is attached to the <u>pituitary gland</u> or the master control gland of the endocrine system.

Midbrain

The midbrain or mesencephalon serves as an important center for the receipt and integration of sensory information. It is particularly involved with visual and auditory processing.

The midbrain is divided by the cerebral aqueduct into a roof (tectum) and floor (cerebral peduncles). The tectum consists of the <u>superior</u> and <u>inferior colliculi</u>. In order to best observe the colliculi look at the brain half with the cerebrum removed and hippocampus exposed. The two prominent bumps are then viewed from either a sagittal or dorsal view. (See **Figure 9-6**.) The larger, more anterior, and superior bump is called the superior colliculus. It coordinates visual reflexes and processes some perceptual functions.

The swollen, more posterior, and inferior bump is called the <u>inferior colliculus</u>. All the nerve and fibers involved in hearing terminate in or pass through the inferior colliculus.

The <u>cerebral peduncles</u> can be seen in the sagittal section although they are best viewed from the ventral side. The cerebral peduncles control states of arousal.

Hindbrain

The hindbrain consists of the pons, cerebellum and the medulla.

Although the <u>pons</u> is always diagrammed as a distinctive bump found on the ventral side, it may be difficult to locate. Its anterior border is well marked by the inferior colliculum. The posterior border is delimited by a slight groove where the cerebellum artery is located. With a sagittal view, this groove appears as a triangle marking the position between the pons and the medulla.

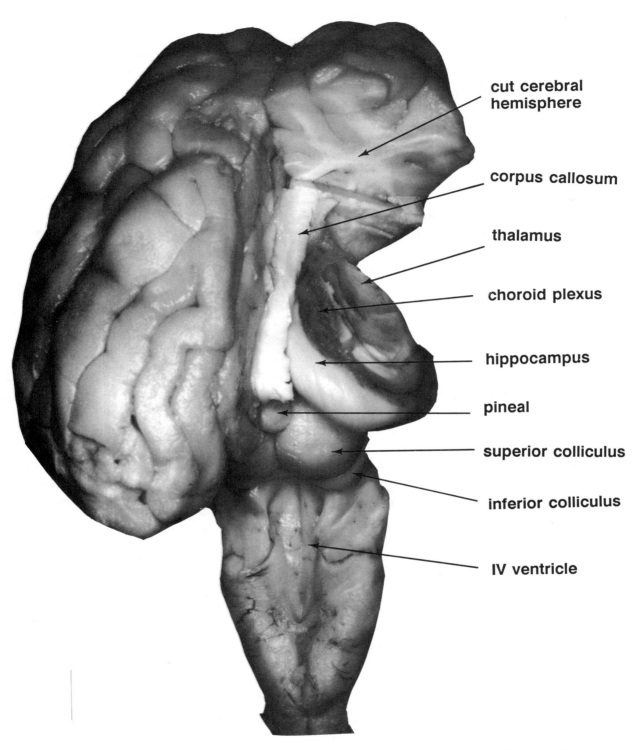

cut cerebral
hemisphere

corpus callosum

thalamus

choroid plexus

hippocampus

pineal

superior colliculus

inferior colliculus

IV ventricle

Figure 9-6. Sheep brain. Dorsal View with Cerebral Hemisphere Removed

The pons contains several fiber tracts and nuclei with almost all sensory and motor information passing through the pons. The two hemispheres of the cerebellum are connected by a fiber tract that goes through the pons. Finally, with the medulla it regulates breathing.

The cerebellum is found on the dorsal side of the cerebral aqueduct. The deep folds are called sulci which are similar to the cerebral cortex. In fact, cerebellum means "little cortex." The cerebellum should be cut in half by making a midsagittal cut. This exposes the distinctive white fiber tract known as the arbor vitae (tree of life).

The cerebellum receives sensory information concerning the position of the joints and lengths of muscles. It then integrates this information with sensory and auditory information to coordinate unconscious movement and maintain balance.

The medulla oblongata lies at the most posterior end of the brain stem. As the motor nerves leave the brain they cross in the medulla at a point known as the pyramidal decussation. The name is derived from the pyramid shaped enlargement of the medulla where the fibers cross. To locate the pyramid turn the brain so that the ventral side is facing you. Then take a probe and stick it between the pons and the medulla. Finally, pull the probe posterially through the medulla. At some point it should get stuck. This point marks where the fibers cross.

The medulla is involved in respiration and states of arousal.

Cross Sections

Sheep brains used for cross sections have been preserved. **Gloves are required.** Exercise extreme caution to avoid open and especially eye contact with this preservative. If preservative is accidentally splashed in the eyes, rinse immediately with copious amounts of clear water.

Each lab bench should obtain a brain. The brain should be rinsed with water for at least 15–30 seconds. To facilitate cutting, any remaining dura mater should be removed from the brain.

Once all of the dura mater has been removed, place the brain on the lab bench ventral side downward. A microtome knife or other large knife should be used to make the cross sectional cuts. The first cut should pass through the optic chiasm. The cuts should be made in a single smooth movement by drawing the knife down and avoiding sawing movements. Make 1 cm thick cuts toward the posterior until the pineal gland is reached.

Staining Procedure

1. Gently agitate sections in tap water in a 4″ finger bowl for about 2 min.

2. Place slices in Mulligan's fluid for 2 min.

3. 1 min wash in tap water.

4. Place slices in 2% aqueous tannic acid for 1 min with gentle agitation.

5. Rinse in tap water for 3 min.

6. Place slices in 1% aqueous iron alum (ammonium ferric sulphate). It is during this step that the gray matter containing the cell bodies becomes stained. The white matter protected by the myelin sheaths remains white. The slices should only be left in the iron alum until the gray tissue is clearly visible, gray or black.

7. Remove the slices and wash in tap water.

In examining the slices it is important to remember that the cut through the optic chiasm produces two slices that are similar to **Figure 9-7.** These slices should be shared among groups working at the same bench. Remember to look at both sides of a slice to determine which side best fulfills your needs. The slice which contains the optic chiasm (**Figure 9-7**) and the hippocampus (**Figure 9-8**) will be required in today's lab.

In examining the slice which contains the optic chiasm several structures previously identified can be viewed in cross sections. In addition, the white matter or fiber tracts become distinct. Fibers can be divided into three groups: association, commisural, and projection.

Association Fibers

Association fibers interconnect cortical regions within the same hemisphere. Several different fibers exist but we will only observe one. The arcuate fibers are short fibers that connect the adjacent gyri. The arcuate fibers are seen in the most dorsal aspect of the cerebrum as U-shaped fibers.

Commisural Fibers

Commisural fibers connect the two hemispheres of the brain together. Therefore, they cross the longitudinal fissure. We will only look at the major commisural fiber, the corpus callosum. In humans, 300 million nerve fibers pass through the corpus callosum. The white glistening corpus callosum should already be familiar to you in the whole brain specimen. In cross section at the level of the optic chiasm it is the only white fiber that crosses the midline. It is ventral to the cingulate gyrus and dorsal to the lateral ventricles. Recall in the previous dissection how the lateral ventricles were displayed only after removing the corpus callosum.

Clinical Note: Fibers

In some patients with severe uncontrollable epilepsy the corpus callosum used to be cut in order to prevent the spread of the seizure from one hemisphere to the other. These "split-brain" patients appear normal except for some highly specific tasks. For example, if an object is placed in the patients left hand the sensory information goes to the right hemisphere (recall fibers cross at the medulla). Unfortunately language processing is only found in the left hemisphere.

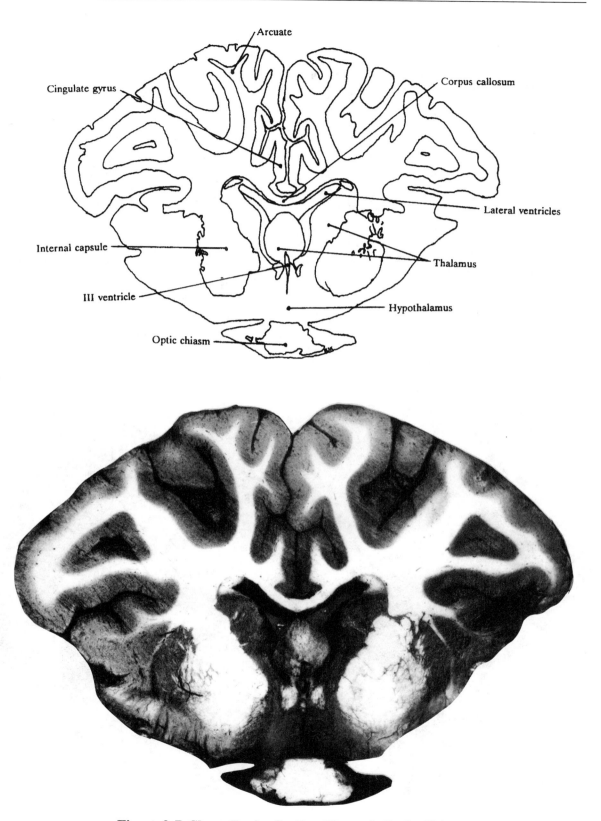

Figure 9-7. Sheep Brain. Section Through Optic Chiasm

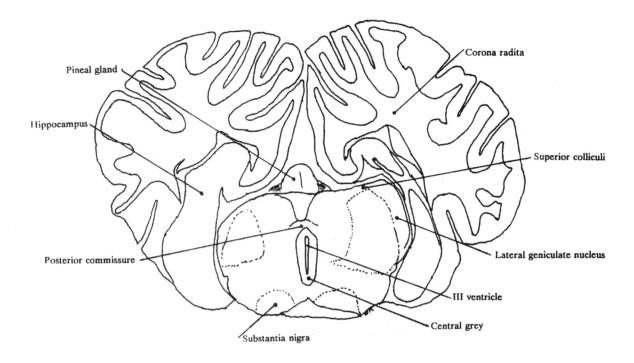

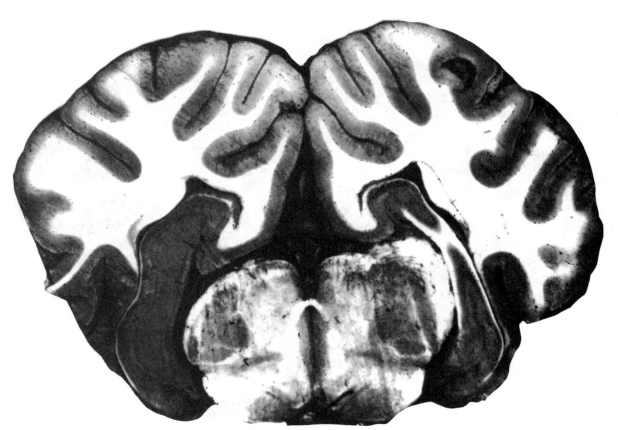

Figure 9-8. Sheep Brain. Section Through Pineal Gland

Therefore, the patient will be unable to name the object. However, the patient has no problem with objects placed in the right hand.

Projection Fibers

Projection fibers connect the cerebral cortex with subcortical regions. We will observe the internal capsule which contains fibers coming and going from the thalamus and cerebral cortex. The internal capsule is a distinctive paired collection of white fibers on the ventral half of the body and lateral to the thalamus.

Other Structures

Observing your slice and **Figure 9-7,** identify and become familiar with the cingulate gyrus which is found deep within the longitudinal fissure. The cingulate gyrus is part of the neocortex that is often associated with emotions. Identify the lateral ventricles which are just ventral of the corpus callosum. Also identify the narrow third ventricle which is exactly in the midline. The third ventricle helps to distinguish the thalamus from the hypothalamus. The thalamus is located above the third ventricle while the hypothamalus is underneath. If the optic chiasm stayed attached to the brain it should also be identified. However, the process of making the initial slices sometimes causes the optic chiasm to be removed.

In the slice that contains the hippocampus (see **Figure 9-7**) several additional structures may be identified. It is important to realize that every brain slice is different and will not precisely match the figures. In some slices it may still be possible to view the corpus callosum. In others the pineal gland may be present. In all of these slices the internal capsule is no longer present. Instead the most prominent white fiber tract is the corona radiata. These projection fibers travel to the cerebral cortex. The hippocampus is seen as a large curved mass of gray matter below the corona radita. If your staining was successful it may be possible to see the three histologically distinct regions of the hippocampus.

The brain stem is quite distinct in the hippocampal section. The third ventricle is distinct and still present in the exact midline. It is now surrounded by gray matter called the central gray. This is the beginning of a distinct feature found in the spinal cord. Dorsal to the central gray is a small commisural fiber connecting the hemispheres called the posterior commissure. The most lateral part of the brain stem is the lateral geniculate nucleus where some visual processing takes place. The most ventral part of the brain stem contains the small gray area called the substantia nigra. When this part of the brain is destroyed Parkington's disease results. After identifying the structures make sure the brain slices are shared among other partners in the class room. Take the opportunity to look at other slices in order to appreciate differences between regions of the brain.

Histology

Slide #1 Myoneural junction. The axons of motor nerves exit the ventral root of the spinal cord and travel to skeletal muscles which they innervate and thereby cause to contract. The axons branch and form terminal dilations on individual muscle fibers. Distinguish between the branched nerve cell axon and the muscle cells (10x). A <u>myoneural junction</u> (also called motor end-plate) can be found by tracing to the distal tip of an axon and locating a dark staining plate-like structure on the surface of a muscle fiber. Between axon and muscle is the synaptic cleft, through which action potentials are transmitted.

Slide #2 Pacinian corpuscle. <u>Pacinian corpuscles</u> are large sensory structures which function to detect changes in pressure (mechanoreceptors). These structures are conspicuous and resemble a sliced onion in histologic sections and are found in the skin (especially fingertips) and mesentery (10x). Pacinian corpuscles consist of 20–70 concentric layers of fibroblast cells (connective tissue cells) alternating with fluid-filled spaces that surround a centrally located nerve terminal. Consider how the structure of the corpuscle relates to its function.

Bio 204 Checklist

Lab 9—Kingdom ANIMALIA

Phylum Chordata

The Brain

<u>Lobes</u>:   frontal
parietal
temporal
occipital
pyriform

<u>Nerves</u>:   I.   olfactory bulb
II.   optic nerve
III.   oculomotor
IV.   trochlear
V.   trigeminal
VI.   abducens
VII.   facial
VIII.   vestibulo-cochlear
IX.   glossopharyngeal
X.   vagus
XI.   spinal accessory
XII.   hypoglossal

Histology:

myoneural junction
Pacinian corpuscles

Structures:

dura mater
pia mater
cerebrum
cerebellum
medulla oblongata
corpus callosum
hippocampus
choroid plexus
septum pellucidum
thalamus
pineal body
hypothalamus
pituitary gland
superior colliculus
inferior colliculus
cerebral peduncles
pons
fissure
sulci
gyri
optic chiasm
ventricles (III, IV)
pyramidal decussation
arbor vitae
corona radiata
fibers: arcuate, commisural,
    association, projection,
    myelinated